HUMAN ANATOMY AND PHYSIOLOGY

ANATOMY AND PHYSIOLOGY RESEARCHER BIOGRAPHICAL SKETCHES AND RESEARCH SUMMARIES

HUMAN ANATOMY AND PHYSIOLOGY

Additional books in this series can be found on Nova's website under the Series tab.

Additional E-books in this series can be found on Nova's website under the e-book tab.

PHYSIOLOGY - LABORATORY AND CLINICAL RESEARCH

Additional books in this series can be found on Nova's website under the Series tab.

Additional E-books in this series can be found on Nova's website under the e-book tab.

HUMAN ANATOMY AND PHYSIOLOGY

ANATOMY AND PHYSIOLOGY RESEARCHER BIOGRAPHICAL SKETCHES AND RESEARCH SUMMARIES

NEIL R. ZIRENA
AND
XIA-BO LIU
EDITORS

NOVA BIOMEDICAL
New York

For permission to use material from this book please contact us:
Telephone 631-231-7269; Fax 631-231-8175
Web Site: http://www.novapublishers.com

Library of Congress Cataloging-in-Publication Data

ISBN: 978-1-61942-595-8

Published by Nova Science Publishers, Inc. ✝ New York

Contents

Preface

This new book compiles biographical sketches of top professionals in the field of anatomy and physiology research, as well as research summaries from a number of different focuses in these important fields.

Part 1 - Research Biographies

In: Anatomy and Physiology Research ...
Editors: N. R. Zirena and X. B. Liu
ISBN: 978-1-62942-595-8

Chapter 1

Biographical Sketch - Anatomy and Physiology Research

Hidekazu Aoyagi

Affiliation:

Advanced Research Center, School of Life Dentistry at Niigata, the Nippon Dental University

Address:

1-8 Hamaura-cho, Chuo-ku, Niigata 951-8580, Japan

Date of Birth:

1950/07/25

Research and Professional Experience:

Histopathology

Professional Appointments:

Assistant professor

Publications Last 3 Years:

1) Iwasaki S, Aoyagi H, Yoshizawa H: Localization of keratins 13 and 14 in the lingual mucosa of rats during the morphogenesis of circumvallate papillae, Acta Histochem, 113:395-401,2011.
2) Iwasaki S, Aoyagi H: Fluorescence immunohistochemistry in combination with differential interference contrast microscopy for studies of semi-ultrathin specimens of epoxy resin-embedded samples (Chapter 14), In: Methods in Molecular Biology Vol. 689, Light Microscopy Methods and Protocols (Eds. H. Charini-Garcia & R. C. N. Melo) pp. 229-240, 2010, Humana Press c/o Springer Science + Business Media, New York, USA.
3) Aoyagi H, Tsuchikawa K, Iwasaki S: Three-dimensional observation of the mouse embryo by micro-computed tomography: composition of the trigeminal ganglion, Odontology, 98: 26-30,2010.
4) Iwasaki S, Asami T, Wanichanon C, Yoshizawa H, Aoyagi H: Immunohistochemical analysis of type III collagen expression in the lingual mucosa of rats during organogenesis of the tongue, Odontology, 96(1): 12-20 2008.
5) Asami T, Aoyagi H, Yoshizawa H , Wanichanon C , Iwasaki S:Immunohistochemical expression of type II collagen in the lingual mucosa of rats during organogenesis of the tongue, Arch Oral Biol, 53(7): 622-8, 2008.
6) Aoyagi H, Asami T, Yoshizawa H , Wanichanon C , Iwasaki S:Newly developed technique for dual localization of keratins 13 and 14 by fluorescence immunohistochemistry, Acta Histochem, 110(4): 324-32, 2008.
7) Hidekazu AOYAGI and Shin-ich IWASAKI:Long-term Effect of Silver Powder in vivo,Dent Mater J 27(4): 612-625,2008.

In: Anatomy and Physiology Research
Editors: N. R. Zirena and X. B. Liu
ISBN: 978-1-62942-595-8

Chapter 2

Biographical Sketch - Anatomy and Physiology Research

Takayuki Baba

Affiliation:

Department of Ophthalmology and Visual Science, Chiba University Graduate School of Medicine, Chiba, Japan

Address:

1-8-1 Inohana, Chuo-ku, Chiba260-8670, Japan

Date of Birth:

7/6/1972

Research and Professional Experience:

Vitreoretinal Diseases

Professional Appointments:

Associate Professor, Chiba University Graduate School of Medicine, 2005-current

Publications Last 3 Years:

1. Mitamura Y, Kubota-Taniai M, Okada K, Kitahashi M, Baba T, Mizunoya S, Yama moto S. Comparison of photodynamic therapy to transpupillary thermotherapy for polypoidal choroidal vasculopathy. Eye 2009; 23: 67-72. Epub 2007 Sep 28.
2. Aizawa S, Mitamura Y, Baba T, Hagiwara A, Ogata K, Yamamoto S. Correlation between visual function and photoreceptor inner/outer segment junction in pati ents with retinitis pigmentosa. Eye 2009; 23: 304-8. Epub 2008 Jan 11.
3. Mitamura Y, Aizawa S, Baba T, Hagiwara A, Yamamoto S. Correlation between R etinal Sensitivity and Photoreceptor Inner/Outer Segment Junction in Patients with Retinitis Pigmentosa. Br J Ophthalmol 2009; 93:126-127.
4. Mitamura Y, Hirano K, Baba T, Yamamoto S. Correlation of Visual Recovery to Presence of Photoreceptor Inner/Outer Segment Junction in Optical Coherence I mages after Epiretinal Membrane Surgery. Br J Ophthalmol 2009; 93: 171-175. Ep ub 2008 Oct 29.
5. Baba T, Grebe R, Hasegawa T, Bhutto I, Merges C, McLeod DS, Lutty GA. Matur ation of the fetal human choriocapillaris. Invest Ophthalmol Vis Sci 2009; 50: 3503-3511. Epub 2009 Mar 5.
6. McLeod DS, Grebe R, Bhutto I, Merges C, Baba T, Lutty GA. Relationship of R PE and choriocapillaris in age-related macular degeneration. Invest Ophthalmol Vis Sci 2009; 50: 4982-4991. Epub 2009 Apr 8.
7. Okada K, Kubota-Taniai M, Kitahashi M, Baba T, Mitamura Y, Yamamoto S. Chan ges in visual function and thickness of macula after photodynamic therapy for age-related macular degeneration. Clinical Ophthalmology 2009; 3: 483-438. Epu b 2009 Sep 7.
8. Bhutto IA, Baba T, Merges C, McLeod DS, Lutty GA. Low nitric oxide synthese s (NOS) in eyes with age-related macular degeneration (AMD). Exp Eye Res 2010; 90: 155-67. Epub 2009 Oct 15.
9. Sengupta N, Afzal A, Caballero S, Chang K, Shaw LC, Pang J, Bond VC, Bhutto I, Baba T, Lutty GA, Grant MB. Paracrine modulation of CXCR4 by IGF-1 and VEG F: Implications for choroidal neovascularization. Invest Ophthalmol Vis Sci 20 10; 51: 2697-704. Epub 2009 Dec 10. doi:10.1167/iovs.09-4137.
10. Baba T, Kubota-Taniai M, Kitahashi M, Okada K, Mitamura Y, Yamamoto S. Two - year comparison of photodynamic therapy and intravitreal bevacizumab for trea tment of myopic choroidal neovascularization. Br J Ophthalmol 2010; 94: 860-70 . Epub 2009 Dec 3.
11. Lutty GA, Hasegawa T, Baba T, Grebe R, Bhutto IA, McLeod DS. Development of the human choriocapillaris. Eye 2010; 24: 408-15. Epub 2010 Jan 15.
12. Baba T, Bhutto IA, Merges C, Grebe R, Emmert D, McLeod DS, Armstrong D, Lutty GA. A rat model for choroidal neovascularization using subretinal lipid hydroperoxide injection. Am J Pathol 2010; 176: 3085-97. Epub 2010 Apr 15.
13. Barcelona$B!!(JPF, Luna JD, Chiabrando GA, Juarez C, Bhutto IA, Baba T, McLeod DS, Sanchez MC, Lutty GA. Immunohistochemical localization of low density lipoprotein receptor-related protein 1 and $B&A(J2-macroglobulin in human retinal and choroidal tissue of proliferative retinopathies. Exp Eye Res 2010; 91: 264-72. Epub 2010. Jun 1.

14. Kajita F, Oshitari T, Yotsukura J, Asanagi K, Baba T, Kishimoto T, Yamamoto S. Case of primary diffuse large B-cell lymphoma of lacrimal sac in a Japanese patient. Clinical Ophthalmology 2010; 4: 1351-4.
15. Oshitari T, Yotsukura J, Asanagi K, Baba T, Kishimoto T, Yamamoto S. Relat ionship between chronic sclerosing dacryoadenitis with high level of IgG4 andCastleman disease. Clinical Ophthalmology 2011; 5: 23-5.
16. Ooka M, Mitamura Y, Baba T, Kitahashi M, Oshitari T, Yamamoto S. Foveal microstructure on spectral-domain optical coherence tomographic images and visual function after macular hole surgery. Am J Ophthalmol Epub 2011. doi: 10.1016/j.ajo.2011.02.001
17. Baba T, Tatsumi T, Oshitari T, Yamamoto S. Case of retinal detachment due to retinal break splitting macula vertically in patient with cognitive disorder. Clinical Ophthalmology 2011; 5: 411-3.
18. Baba T, Kitahashi M, Kubota-Taniai M, Oshitari T, Yamamoto S. Subretinal Hemorrhage after Photodynamic Therapy for Juxtapapillary Retinal Capillary Hemangioma. Case Rep Ophthalmol 2011; 2: 134-9.
19. NizawaT, Oshitari T, Kimoto R, Kajita F, Yotsukura J, Asanagi K, Baba T, Takahashi Y, Oide T, Kiyokawa T, Kishimoto T, Yamamoto S. Early-stage mucinoussweat gland adenocarcinoma of eyelid. Clinical Ophthalmology 2011:5 687?689.
20. Bhutto IA, Baba T, Merges C, Juriasinghani V, McLeod DS, Lutty GA. C-reactive protein and complement factor H in aged human eyes and eyes with age-related macular degeneration. Br J Ophthalmol 2011 on line first.

In: Anatomy and Physiology Research
Editors: N. R. Zirena and X. B. Liu

ISBN: 978-1-62942-595-8

Chapter 3

Biographical Sketch - Anatomy and Physiology Research

Rubens Cecchini

Affiliation:

Londrina State University

Address:

Rodovia Celso Garcia Cid, s/n, km 380, PR-445, Bairro Versalles, Londrina, PR, Brazil, CEP 86051-980

Date of Birth:

1946

Research and Professional Experience:

Postdoctoral Fellow (1988-1990) at the University of London King's College in the area of Free Radicals In Medicine. He is currently an associate professor atState University of Londrina. He has experience in medical research, with emphasis on Pathophysiology and Experimental Pathology, acting on the following topics: pre-hemolytic oxidative lesions, involvement of ROS and RNS damage in ischemia /reperfusion of skeletal muscle, oxidative stress in cachexia, oxidative injury UV-induced skin. Oxidative stress and breast cancer.

Publications Last 3 Years:

Pavanelli, Wander Rogério, Silva, Jean Jerley Nogueira, Panis, Carolina, Cunha, Thiago Mattar, Costa, Ivete Conchon, Menezes, Maria Claudia Noronha Dutra, Abreu Oliveira, Francisco José, França Lopes, Luiz Gonzaga, Cecchini, Rubens, Queiroz Cunha, Fernando, Watanabe, Maria Angélica Ehara, Itano, Eiko Nakagawa Experimental Chemotherapy in Paracoccidioidomycosis Using Ruthenium NO Donor. Mycopathologia (1975. Print). , v.172, p.95 - 107, 2011.

Simão, A.N.C., Lozovoy, M.A.B., Simão, T.N.C., Venturini, D., Barbosa, D.S., Dichi, J.B., Matsuo, T., Cecchini, R., Dichi, I.

Immunological and biochemical parameters of patients with metabolic syndrome and the participation of oxidative and nitroactive stress. brazilian journal of medical and biological research -on line. , v.44, p.707 - 712, 2011.

Panis, C., Herrera, A. C. S. A., Victorino, V. J., Campos, F. C., Freitas, L. F., Rossi, T., Colado Simão, A. N., Cecchini, A. L., Cecchini, R.

Oxidative stress and hematological profiles of advanced breast cancer patients subjected to paclitaxel or doxorubicin chemotherapy. Breast Cancer Research and Treatment. , v.xxx, p.xx - xx, 2011.

Lozovoy, M. A. B., Simao, A. N. C., Panis, C., Rotter, M. A. C., Morimoto, H. K., Reiche, E. M. V., Lavado, E., Cecchini, R., Dichi, I.

Oxidative stress is associated with liver damage, inflammatory status, and corticosteroid therapy in patients with systemic lupus erithematosus. Lupus (Basingstoke). v.xxx, p.xx - xx, 2011.

Peres, P.S., Terra, V.A., Guarnier, F.A., Cecchini, R., Cecchini, A.L.

Photoaging and chronological aging profile: Understanding oxidation of the skin. Journal of Photochemistry and Photobiology. B, Biology. , v.103, p.93 - 97, 2011.

Panis, C., Mazzuco, T. L., Costa, Cauê Zortéa Fernandes, Victorino, V. J., Tatakihara, V. L. H., Yamauchi, L. M., Yamada-Ogatta, S.F., Cecchini, R., Rizzo, L. V., Pinge-Filho, P.

Trypanosoma cruzi: Effect of the absence of 5-lipoxygenase (5-LO)-derived leukotrienes on levels of cytokines, nitric oxide and iNOS expression in cardiac tissue in the acute phase of infection in mice.. Experimental Parasitology. , v.127, p.58 - 65, 2011. Cunha,N.V., Abreu, S.B., PANIS, C., Grassiolli, S.

Guarnier, Flávia Alessandra, Cecchini, R., Mazzuco, T. L., Pinge Filho, P., Martins-Pinge, M. C. Cox-2 inhibition attenuates cardiovascular and inflammatory aspects in monosodium glutamate-induced obese rats. Life Sciences (1973). , v.87, p.375 - 381, 2010.

Guarnier, Flávia A., Cecchini, Alessandra L., Suzukawa, Andréia A., Maragno, Ana Leticia G.C., Simão, Andréa N.C., Gomes, Marcelo D., Cecchini, Rubens Time course of skeletal muscle loss and oxidative stress in rats with walker 256 solid tumor. Muscle & Nerve (Print). , v.42, p.950 - 958, 2010.

De Rossi, T., Panis, C., Victorino, V. J., Freitas, L. F., Herrera, A. C. S. A., Cecchini, A. L., Cecchini, R. Breast Cancer and Oxidative Stress in chemotherapy. Applied Cancer Research (Impresso). , v.29, p.150 - 156, 2009. Bregano, J W, Dichi JB, Barbosa, D. S., Kadri, M. Z. E., Matsuo T, Rodríguez, M. A. M., Cecchini, R., Dichi, I. Decreased Total Antioxidant Capacity in Plasma, but Not Tissue, in Experimental Colitis. Digestive Diseases and Sciences. , v.54, p.751 - 757, 2009.

Borges, C.L., Cecchini, R., Tatakihara, V. L. H., Malvezi, A. D., Yamada-Ogatta, S.F., Rizzo, L. V., Pinge Filho, P.

5-Lipoxygenase plays a role in the control of parasite burden and contributes to oxidative damage of erythrocytes in murine Chagas' disease. Immunology Letters. , v.123, p.38 - 45, 200.

In: Anatomy and Physiology Research
Editors: N. R. Zirena and X. B. Liu
ISBN: 978-1-62942-595-8

Chapter 4

Biographical Sketch - Anatomy and Physiology Research

Haibin Chen

Affiliation:

Associated Research Professor of Biomechanics

Address:

Head/Neck/Lower extremity impact biomechanics

Date of Birth:

11-July-1965

Research and Professional Experience:

Dr CHEN is an Associated Research Professor at Daping Hospital of the Third Military Medical University. During 2007-2008, Dr. CHEN had been supported by the Chinese government to work at the Bioengineering Center of Wayne State University in USA as a 13-month visiting scholar and supervised by Professor Albert I KING (Dr King is one of the most famous scientists of impact biomechanics in the world). His areas of expertise include impact biomechanics, blood rheology, injury models and precision instrument. From 2004 he has been an engineering senior specialist at Crash Test Division of State Key Laboratory of Trauma, Burns, and Combined Injury at the Third Military Medical University. His activities are focused on vehicle crash tests, and head/neck/lower extremity impact biomechanics.

Professional Appointments:

Dr. Haibin CHEN has been an executive council member of the Biomedical Engineering Society of Chongqing, and a member of the International Traffic Medicine Association (ITMA) and the Society of Automotive Engineers (SAE) as well as the American Society of Mechanical Engineers (ASME).

Honors:

The forth leading winner of a China National Second Prize of Science and Technology Progress.

Publications Last 3 Years:

Haibin Chen, Zhengguo Wang, Kai Xiao, et al. X-ray observation on how axial compression stimulates tibial fracture healing. Chin J Traumatol 2010; 13(6): 323-328.

Haibin Chen, Jianjin Hang, Bo Zhang, et al. Establishment of the model of motorcyclist ejection injury. Chin J Traumatol 2010; 13(2): 67-71.

Haibin Chen, Zhengguo Wang, Albert I King, et al. Injury tolerance and injury mechanism of lower extremity in automobile impact accidents. Journal of automotive safety and energy 2010, 1(4): 1-10. *in Chinese.*

Haibin Chen, King H Yang, Zhengguo Wang. Biomechanics of whiplash injury. Chin J Traumatol 2009; 12(5): 305-314.

Haibin Chen, Xuemei Cheng, King H Yang. Two-dimensional numerical simulations of a biphasic representation of human tibia in car-pedestrian accidents. ASME Paper No. IMECE2008-69307: 1-12. Oct 31 – Nov 6, 2008; Bostoon, USA.

Haibin Chen. Injury patterns and risk factors of motorcycle crashes. Chinese Journal of Clinical Rehabilitation. 2006, 10(40): 187-191.

Haibin Chen, Zhengguo Wang, Kai Xiao, et al. X-ray observation of tibial fracture healing due to mechanical stress stimulation. The Proceeding of 7th Int. Forum of Automotive traffic Safety (INFATS). Dec 4-6, 2009; Changsha, PR China.

Haibin Chen, Liying Zhang, Zhengguo Wang, et al. Clinical and biomechanics analyses of whiplash injury. The Proceeding of 8th Int. Forum of Automotive traffic Safety (INFATS). Dec 2-4, 2010; Wuhu, PR China.

In: Anatomy and Physiology Research
Editors: N. R. Zirena and X. B. Liu

ISBN: 978-1-62942-595-8

Chapter 5

Biographical Sketch - Anatomy and Physiology Research

Frédérique Courtois

Affiliation:

Université du Québec à Montréal, Dept of Sexology

Address:

Université du Québec à Montréal, Dept of Sexology, CP 8888 succursale centre villle, Montreal, Quebec, H3C 3P8 Canada

Date of Birth:

September 24, 1957

Education:

PhD and post doctorate fellowship

Research and Professional Experience:

23 years experience in research and rehabilitation

Professional Appointments:

Clinical consultant for sexual dysfunctions in men and women with SCI at 1) Institut de réadaptation Gingras Lindsay de Montréal, 2) Institut de réadaptation en déficience physique de Québec since 1991 and 1994 respectively

Honors:

Deans list of honors for PhD thesis, Post doctoral research fellows in 1989 and University research fellow in 1991 and 1994

Publications Last 3 Years:

Book Chapters

Courtois F., Charvier K, Vézina J-G, Carrier S, Morel Journel N, Jacquemin G, Leriche A. Assessment of sexual potential and treatment of sexual dysfunctions in men and women with spinal cord injury. Ed T C Berkovsky Handbook of spinal cord injuries: Types, treatments and progronis. New York: Novapublishers. 2009.

Courtois, F.. Réversibilité des facteurs de risque et impact des mesures hygiéno-diététiques dans la DE. Dans Dysfonction érectile et maladies cardio-vasculaires, Ed. B Cuzin. Collection Pathologie Science Formation. John Libbey Eurotext. 2009.

Courtois, F, Mathieu, C & Bélanger D & Paradis, M. Lésions médullaires. Dans Dictionnaire de la sexualité au Québec. Eds Levy, J.J & Dupras, A. 2009.

Articles in refereed Journals:

Otis, C, Courtois, F & Marchand, A. Risk factors for posttraumatic stress disorder in the spinal cord injured population. Under revision.

Otis, C, Courtois, F & Marchand, A. Posttraumatic stress disorder in the spinal cord injured population : a review of the literature. Under revision.

Caremel R, Courtois F, Charvier K, Ruffion A, Morel Journel N. Side effects of intradetrusor botulic toxin injections on ejaculation and fertility in men with spinal cord injury. British Journal of Urology, 2011, DOI:10.1111/j.1464-410X.2011.10639x.

Courtois, F, Charvier, K., Vézina, J.-G., Morel-Journel N, Carrier S., Jacquemin, G, Côté, I. Assessing and conceptualizing orgasm following a spinal cord injury. British Journal of Urology, 2011, DOI.10.1111/j.1464-410X.2011.10168x.

Morel-Journel N, Paparel, P, Ruffion, A, Courtois F, Leriche A. Maladie de la Peyronie : Prise en charge en 2009. Progrès en Urologie (Formation Médicale Continue).

Courtois F. Réhabilitation des traumatismes sexuels : une priorité clinique/Rehabilitation of sexual trauma : a clinical priority. European Journal of Sexology and Sexual Health, 2009 18, 77-78 (angl) et 112-113 (fr).

Courtois F, Charvier K., Leriche, A., Vézina JG, Jacquemin G. Réactions sexuelles et orgasme chez le patient traumatisé médullaire: un modèle de rééducation/ réadaptation/

Sexual and climactic responses in men with traumatic spinal injury: A model for rehabilitation. European Journal of Sexology and Sexual Health, 2009 18, 79-82 (angl)et 114-120 (fr).

Morel-Journel N, Courtois F, Paparel, P, Ruffion, A, Leriche A. Reconstruction chirurgicale après traumatismes sexuels et malformations sexuelles congénitales majeures /Reconstructive surgery for major sexual anomalies in adults. European Journal of Sexology and Sexual Health 2009 18, 98-104 (angl)et 147-155 (fr).

In: Anatomy and Physiology Research
Editors: N. R. Zirena and X. B. Liu

ISBN: 978-1-62942-595-8

Chapter 6

Biographical Sketch - Anatomy and Physiology Research

Vladimir Danilovich Kreslavski

Affiliation:

Institute of Basic Biological Problems,
Russian Academy of Sciences.
Pushchino, Moscow Region, Russia

Address:

Institute of Basic Biological Problems,
Russian Academy of Sciences,
St. Institutskaya, 2. Pushchino, Moscow Region,
142290, Russia.
Phone (w): (496)7330532; 732988;
Phone (h): (496)7734884;
Fax: (496)7330532
E-mail address: vkreslav@rambler.ru

Date of Birth:

17.03. 1948.

Education:

1971: University diploma; photonica; Obtained from Leningrad State University, Russia.

1983: Ph.D; biophysics; Institute of Biophysics, Russian Academy of Sciences, Pushchino.

2011: Doctor of sciences; plant physiology and biochemistry; Institute of Plant Physiology, Russian Academy of Sciences, Moscow.

Research and Professional Experience:

Molecular mechanisms of stress resistance of photosynthetic machinery, stress and light signaling, photoreceptors.

Professional Appointments:

At the present senior researcher, head of group at the Institute of Basic Biological Problems, Pushchino, Russia.

Honors:

Pending of Soros International Science Foundation, Russia, 1994.

Publications Last 3 Years:

Kreslavski V. D., Carpentier R., Klimov V. V., Murata N., Allakhverdiev S. I. Molecular Mechanisms of stress resistance of the photosynthetic apparatus // Biochemistry (Moscow) Supplement Series A: Membrane and Cell Biology. 2007. V. 1(3). P. 185–205.

Kreslavski V., Tatarinzev N., Shabnova N., Semenova G., Kosobryukhov A. Characterization of the nature of photosynthetic recovery of wheat seedlings from short-time dark heat exposures and analysis of the mode of acclimation to different light intensities // J. Plant Physiol. 2008. V. 165. P. 1592-1600.

Allakhverdiev S.I., Kreslavski V.D., Klimov V.V., Los D.A., Carpentier R., Mohanty P. Heat stress: An overview of molecular responses in photosynthesis // Photosynthesis Research. 2008. V. 98(1-3). P. 541-550.

Kreslavski V.D., Lubimov V.Yu., Shabnova N.I., Balakhnina T.I., Kosobryukhov A.A. Heat-induced impairments and recovery of photosynthetic machinery in wheat seedlings. Role of light and prooxidant-antioxidant balance // Physiol. Mol. Biol. Plants. 2009. V.15(2). P. 115-122.

Kreslavski V.D., Carpentier R., Klimov V.V., Allakhverdiev S.I. Transduction mechanisms of photoreceptor signals in plant cells // J. Photochem. Photobiol. C. Photochem. Reviews. 2009 V.10. P. 63–80.

Kreslavski V., Fomina I., Ivanov A.,Tatarinzev N., Kosobryukhov A., Biel K., Herbert S. NaCl-induced photoinhibition and the recovery of the photosynthetic activity of *katG*$^-$ mutant of the cyanobacterium *Synechocystis* sp. PCC 6803. Biophysics. V. 55(2). 2010. P. 215-220.

Kreslavski V. D., Balakhnina T. I., Zharmukhamedov S. K., Shabnova N. I., Khristin M. S., Lyubimov V. Yu. Mechanism of enhancing photosystem 2 thermoresistance in wheat plants by chlorocholine chloride // Russian Agricultural Sciences. 2010. V. 36(3). P. 156–159.

Allakhverdiev S.I., Kreslavski V.D., Fomina I.R., Los D.A., Klimov V.V., Mimuro M., Mohanty P., Carpentier R. Inactivation and repair of photosynthetic machinery under heat stress. In: Photosynthesis: Overviews on Recent Progress & Future Perspective (Eds. Guruprasad K.N., Itoh S., Mohanty P.). – Narosa Publishing House, New Delhi, 2010, Chapter 14, p. 187-214.

Kreslavskii V. D., Lubimov V. Yu., Kotova L. M., Kotov A. A. Effect of common bean seedling pretreatment with chlorocholine chloride on photosystem II tolerance to UV-B radiation, phytohormone content, and hydrogen peroxide content // Russ. J. Plant Physiol. 2011, V. 58(2). P. 324–329.

In: Anatomy and Physiology Research
Editors: N. R. Zirena and X. B. Liu

ISBN: 978-1-62942-595-8

Chapter 7

Biographical Sketch - Anatomy and Physiology Research

Fadi Dornaika

Affiliation:

University of the Basque Country, IKERBASQUE, Basque Foundation for Science

Address:

University of the Basque Country, UPV-EHU, Manuel Lardizabal, 1, 20018 San Sebastian , Spain

Date of Birth:

November 19, 1967

Education:

PhD in signal, image, and speech processing from INRIA, Grenoble, France

Research and Professional Experience:

20 years of research in the field of computer vision and pattern recognition with a focus on geometrical and statistical modeling

Professional Appointments:

Researcher at several research institutes including INRIA, France, University of Alberta, Canada, University of Linkoping, Sweden, and Computer Vision Center, Spain.

Publications Last 3 Years:

1. F. Dornaika, A. Assoum and A. Moujahid. Enhanced and parameterless Locality Preserving Projections for face recognition. Ambient Intelligence and Smart Environments, (Book chapter), IOS press, 2011.
2. F. Dornaika and B. Raducanu. *Subtle facial expression recognition in still images and videos.* (Book chapter) Advances in Face Image Analysis: Techniques and Technologies, Zhang Yu-Jin, editor, IGI Global, 2011.
3. F. Dornaika and J. Elder. Image registration for foveated panoramic sensing. ACM Transactions on Multimedia Computing, Communications, and Applications. In press. (main author).
4. F. Dornaika, J. M. Alvarez, A. D. Sappa, and A. Lopez. A new framework for stereo sensor pose through road segmentation and registration. IEEE Transactions on Intelligent Transportation Systems. To appear 2011.
5. K. Hammoudi and F. Dornaika. A featureless approach to 3D polyhedral building modeling from aerial images. Sensors Journal, volume 11, number 1, pp. 228-259, January 2011.
6. F. Dornaika and B. Raducanu. *3D face pose detection and tracking using monocular videos: tool and application.* IEEE Transactions on Systems, Man, and Cybernetics, Part B, volume 39, number 4, pp. 935-944, August 2009.
7. F. Dornaika and A. Sappa. *Real-time vehicle ego-motion using stereo pairs and particle filters.* Image and Vision Computing, volume 27, number 9, pp. 1382-1393, August 2009.
8. F. Dornaika and A. Sappa. *Instantaneous 3D motion from image derivatives using the least trimmed square regression.* Pattern Recognition Letters, volume 30, number 5, pp. 535-543, April 2009.
9. F. Dornaika and F. Davoine. *Simultaneous facial action tracking and expression recognition in the presence of head motion.* International Journal of Computer Vision, volume 76, number 3, pp. 257-281, March 2008.

In: Anatomy and Physiology Research
Editors: N. R. Zirena and X. B. Liu

ISBN: 978-1-62942-595-8

Chapter 8

Biographical Sketch - Anatomy and Physiology Research

Bernard Geny

Affiliation:

Service de Physiologie et d'Explorations Fonctionnelles et Equipe d'Accueil 3072

Address:

Service de Physiologie et d'Explorations Fonctionnelles,
Pôle de Pathologie Thoracique, Hôpitaux Universitaires,
CHRU de Strasbourg, 67000 Cedex, France.
E-mail: Bernard.geny@chru-strasbourg.fr

Date of Birth:

29.09.1959

Education:

Université et Hôpitaux Universitaires de Strasbourg; MD, PhD;
Phone: 0033369550879.
Fax: 0033369551826.
67091 Strasbourg, France.

Research and Professional Experience:

Cardiology and Physiology. Research on brain natriuretic peptide, heart transplantation, skeletal muscle, ischemia-reperfusion, mitochondria

Professional Appointments:

Professor

Publications Last 3 Years:

1. Bouitbir J, Charles AL, Rasseneur L, Dufour SP, Piquard F, Geny B, Zoll J. Atorvastatin treatment reduces exercise capacities in rats: Involvement of mitochondrial impairments and oxidative stress. J Appl Physiol. 2011 Aug 18.
2. Bouitbir J, Charles AL, Echaniz-Laguna A, Kindo M, Daussin F, Auwerx J, Piquard F, Geny B, Zoll J. Opposite effects of statins on mitochondria of cardiac and skeletal muscles: a 'mitohormesis' mechanism involving reactive oxygen species and PGC-1.Eur Heart J. 2011 Jul 20.
3. Daussin FN, Rasseneur L, Bouitbir J, Lang AL, Dufour SP, Geny B, Burelle Y, Richard R. Different Timing of Changes in Mitochondrial Functions following Endurance Training. Med Sci Sports Exerc. 2011 Jun 28.
4. Charles AL, Guilbert AS, Bouitbir J, Goette-Di Marco P, Enache I, Zoll J, Piquard F, Geny B. Effect of postconditioning on mitochondrial dysfunction in experimental aortic cross-clamping. Br J Surg. 2011 Apr;98(4):511-6.
5. Dal-Ros S, Zoll J, Lang AL, Auger C, Keller N, Bronner C, Geny B, Schini-Kerth VB. Chronic intake of red wine polyphenols by young rats prevents aging-induced endothelial dysfunction and decline in physical performance: role of NADPH oxidase. Biochem Biophys Res Commun. 2011 Jan 14;404(2):743-9.
6. Talha S, Charloux A, Enache I, Piquard F, Geny B. Mechanisms involved in increased plasma brain natriuretic peptide after heart transplantation. Cardiovasc Res. 2011 Feb 1;89(2):273-81.
7. Enache I, Oswald-Mammosser M, Scarfone S, Simon C, Schlienger JL, Geny B, Charloux A. Impact of altered alveolar volume on the diffusing capacity of the lung for carbon monoxide in obesity. Respiration. 2011;81(3):217-22.
8. Schlotterbeck H, Greib N, Dow WA, Schaeffer R, Geny B, Diemunsch PA. Changes in Core Temperature During Peritoneal Insufflation: Comparison of Two CO(2) Humidification Devices in Pigs. J Surg Res. 2010 Nov 12.
9. Duteil D, Chambon C, Ali F, Malivindi R, Zoll J, Kato S, Geny B, Chambon P, Metzger D. The transcriptional coregulators TIF2 and SRC-1 regulate energy homeostasis by modulating mitochondrial respiration in skeletal muscles. Cell Metab. 2010 Nov 3;12(5):496-508.
10. Doutreleau S, Rouyer O, Di Marco P, Lonsdorfer E, Richard R, Piquard F, Geny B. L-arginine supplementation improves exercise capacity after a heart transplant. Am J Clin Nutr. 2010 May;91(5):1261-7.

11. Lejay A, Georg Y, Bajcz C, Thaveau F, Geny B, Kretz JG, Chakfé N. Endovascular treatment of infra-popliteal arteries in patients with critical limb ischemia. Acta Chir Belg. 2009 Nov-Dec;109(6):684-93.
12. Goette-Di Marco P, Talha S, Enache I, Weiller MA, Charloux A, Massard G, Kessler R, Piquard F, Geny B. Endocrine heart after lung transplantation: increased brain natriuretic peptide is related to right ventricular function. Transpl Int. 2010 Jul;23(7):728-35.
13. Ponsot E, Dufour SP, Doutreleau S, Lonsdorfer-Wolf E, Lampert E, Piquard F, Geny B, Mettauer B, Ventura-Clapier R, Richard R. Impairment of maximal aerobic power with moderate hypoxia in endurance athletes: do skeletal muscle mitochondria play a role? Am J Physiol Regul Integr Comp Physiol. 2010 Mar;298(3):R558-66.
14. Thaveau F, Zoll J, Bouitbir J, N'guessan B, Plobner P, Chakfe N, Kretz JG, Richard R, Piquard F, Geny B. Effect of chronic pre-treatment with angiotensin converting enzyme inhibition on skeletal muscle mitochondrial recovery after ischemia/reperfusion. Fundam Clin Pharmacol. 2010 Jun;24(3):333-40.
15. Charloux A, Enache I, Richard R, Oswald-Mammosser M, Lonsdorfer-Wolf E, Piquard F, Geny B. Diffusing capacity of the lung for CO and pulmonary blood flow during incremental and intermittent exercise. Scand J Med Sci Sports. 2010 Feb;20(1):e121-9.
16. Rouyer O, Talha S, Di Marco P, Ellero B, Doutreleau S, Diemunsch P, Piquard F, Geny B. Lack of endothelial dysfunction in patients under tacrolimus after orthotopic liver transplantation. Clin Transplant. 2009 Nov-Dec;23(6):897-903.
17. Talha S, Cohen C, Kindo M, Asmane I, Faure A, Eisenmann B, Geny B, Kurtz JE. Emergency management of a particular massive pulmonary embolism.. J Card Surg. 2009 Jul-Aug;24(4):472-4.
18. Thaveau F, Zoll J, Bouitbir J, Ribera F, Di Marco P, Chakfe N, Kretz JG, Piquard F, Geny B. Contralateral leg as a control during skeletal muscle ischemia-reperfusion.J Surg Res. 2009 Jul;155(1):65-9.
19. Oswald-Mammosser M, Charloux A, Enache I, Lonsdorfer-Wolf E, Geny B. A comparison of four algorithms for the measurement of interrupter respiratory resistance in adults.Respir Med. 2009 May;103(5):729-35.
20. Greib N, Schlotterbeck H, Dow WA, Joshi GP, Geny B, Diemunsch PA. An evaluation of gas humidifying devices as a means of intraperitoneal local anesthetic administration for laparoscopic surgery. Anesth Analg. 2008 Aug;107(2):549-51.
21. Daussin FN, Zoll J, Dufour SP, Ponsot E, Lonsdorfer-Wolf E, Doutreleau S, Mettauer B, Piquard F, Geny B, Richard R. Effect of interval versus continuous training on cardiorespiratory and mitochondrial functions: relationship to aerobic performance improvements in sedentary subjects. Am J Physiol Regul Integr Comp Physiol. 2008 Jul;295(1):R264-72.
22. Talha S, Di Marco P, Doutreleau S, Rouyer O, Piquard F, Geny B. Does circulating BNP normalize after heart transplantation in patients with normal hemodynamic and right and left heart functions? Clin Transplant. 2008 Sep-Oct;22(5):542-8.
23. Daussin FN, Zoll J, Ponsot E, Dufour SP, Doutreleau S, Lonsdorfer E, Ventura-Clapier R, Mettauer B, Piquard F, Geny B, Richard R. Training at high exercise intensity promotes qualitative adaptations of mitochondrial function in human skeletal muscle. J Appl Physiol. 2008 May;104(5):1436-41.

24. Charloux A, Chaouat A, Brandenberger G, Piquard F, Weitzenblum E, Geny B. Spontaneous short-term variations of circulating endothelin-1 in pulmonary hypertension. Transl Res. 2008 Mar;151(3):119-21.
25. Geny B, Richard R, Zoll J, Charloux A, Piquard F. Last Word on Point:Counterpoint: Cardiac denervation does/does not play a major role in exercise limitation after heart transplantation. J Appl Physiol. 2008 Feb;104(2):568.
26. Richard R, Zoll J, Mettauer B, Piquard F, Geny B. Counterpoint: Cardiac denervation does not play a major role in exercise limitation after heart transplantation. J Appl Physiol. 2008 Feb;104(2):560-2, discussion 562-4.
27. Charloux A, Brandenberger G, Piquard F, Geny B. Dysregulation of pulsatility in aging IV. Pulsatile signaling and cardiovascular aging: functions and regulation of natriuretic peptide signaling. Ageing Res Rev. 2008 Jul;7(3):151-63.

In: Anatomy and Physiology Research
Editors: N. R. Zirena and X. B. Liu

ISBN: 978-1-62942-595-8

Chapter 9

Biographical Sketch - Anatomy and Physiology Research

Gerrit Grieb

Affiliation:

Plastic Surgery and Hand Surgery,
Burn Center
Medical Faculty, RWTH Aachen University

Address:

Pauwelsstrasse 30, 52074 Aachen, Germany

Date of Birth:

08th of February, 1978

Education:

RWTH Aachen University, Germany (MD 2005); MD
University of Marburg, Germany
University of Santiago de Compostela, Spain

Research and Professional Experience:

Resident in Plastic Surgery, RWTH Aachen University, Germany

Research Fellow, Department of Internal Medicine, Medical School, Yale University, CT USA,
Research Fellow, Institute of Biochemistry, RWTH Aachen University, Germany

Honors:

2005 16th European Students Conference (ESC)
Best oral presentation in surgery
2008 Robert Bosch Foundation
Grant for an initiating project
â€œSino-german network for developing artificial skin substitutesâ€

Publications during Last Three Years:

Grieb G., Piatkowski A., Simons D., HÃrmann N., Dewor M., Steffens G., Bernhagen J., Pallua N. â€œMacrophage migration inhibitory factor is a potential inducer of endothelial progenitor cell mobilization after flap surgeryâ€• . Surgery, 2010, Epub ahead of print.

Grieb G., Merk M., Bernhagen J., Bucala R. â€œMacrophage migration inhibitory factor (MIF)- A promising biomarkerâ€• . Drug News Perspect, 2010 May; 23(4):257-64.

Grottke O., Braunschweig T., Daheim N., Coburn M., Grieb G., Rossaint R., Tolba R. â€œEffect of TachoSil in a Coagulopathic Pig Model with Blunt Liver Injuriesâ€• . J Surg Res, 2010 Mar; Epub ahead of print.

Grieb G., Simons D., Schmitz L., Piatkowski A., Grottke O., Pallua N.â€œGlasgow Coma Scale and laboratory markers are superior to COHb in predicting CO intoxication severetiyâ€• . Burns, 2010 Apr, Epub ahead of print.

Bozkurt A.*, Grieb G.*, OÂ´Dey D., Fuchs P., Waberski T., Krombach G., Pallua N.*contributed equally to this paper, â€œCommon peroneal nerve compression and heterotopic ossification resulting from severe burn injury: a case reportâ€• . J Bone Joint Surg Am, 2010 Apr;92(4):978-83.

Simons D., Grieb G., Hristov M., Pallua N., Weber C., Bernhagen J., Steffens G.â€œHypoxia-induced endothelial secretion of macrophage migration inhibitory factor and role in endothelial progenitor cell recruitmentâ€• . J Cell Mol Med, 2010 Feb, Epub ahead of print.

Piatkowski A. Ulrich D, Grieb G., Pallua N.â€œA new tool for the early diagnosis of carbon monoxide intoxicationâ€• .Inhalation Toxicology, 2009 Nov;21(13):1144-7.

Groger A.*, Piatkowski A.*, Grieb G.*, Wolter T., Fuchs P., Pallua N.*contributed equally to this paper â€œThe mobilisation of mononuclear cells and endothelial progenitor cells after burn injury in a porcine modelâ€• . Burns, 2010 Jun; 36(4):545-51.

Grieb G., Groger A., Piatkowski A., Markowicz M., Steffens G., Pallua N.â€œTissue substitutes with improved angiogenic capabilities: an in vitro investigation with endothelial and endothelial progenitor cellsâ€• . Cells Tissues Organs, 2010;1 91(2):96-104.

Grieb G., Simons D., Piatkowski A., Bernhagen J., Steffens G., Pallua N.â€œMacrophage migration inhibitory factor- A potential diagnostic tool in severe burn injuries?â€• . Burns, 2010 May; 36(3):335-42.

Grieb G., Groger A., Bozkurt A., Stoffels I., Piatkowski A., Pallua N.â€œThe diversity of carbon monoxide intoxication: medical courses can differ extremely- A Case Reportâ€• .Inhalation Toxicology, 2008 Aug;20(19):911-5.

Reviewer for Journal of Biomaterials Applications Diabetes (American Diabetes Association) Molecular Medicine

In: Anatomy and Physiology Research
Editors: N. R. Zirena and X. B. Liu
ISBN: 978-1-62942-595-8

Chapter 10

Biographical Sketch - Anatomy and Physiology Research

Mozammil Hussain

Affiliation:

Ph.D.

Address:

Research Instructor, Division of Research, Logan University, Chesterfield, MO 63017,
Ph: 636-230-1955. Fax: 636-207-2417. E-mail: mozammil.hussain@logan.edu

Education:

2007 - Ph.D; Bioengineering; University of Illinois at Chicago
2002 - M.S.; Mechanical and Aerospace Engineering; Illinois Institute of Technology, Chicago
2000* - M.E; *Degree was awarded in 2004. Mechanical Engineering; Panjab University, India
1998 - B.E.; Mechanical Engineering; North Gujarat University, India

Research and Professional Experience:

2010 – Present Research Instructor

- Division of Research, Logan University, Chesterfield, Missouri

2007 – 2010 Research Associate

- Division of Research, Logan University, Chesterfield, Missouri

2003 – 2007 Research Assistant, University of Illinois at Chicago

- Orthopedic Biomechanics Laboratory, Department of Orthopedic Surgery, Rush University Medical Center, Chicago, Illinois
- Mentor: Raghu N. Natarajan, Ph.D.

2002 Student Engineering Assistant II

- Orthopedic Biomechanics Laboratory, Department of Orthopedic Surgery, Rush University Medical Center, Chicago, Illinois
- Mentors: Raghu N. Natarajan, Ph.D. and Markus A. Wimmer, Ph.D.

1999 – 2000 Teaching Assistant

- Department of Mechanical Engineering, Punjab Engineering College, Chandigarh, India

Committee Assignments

2009 – Present Member, Learning Resources Center Committee, Logan University, Chesterfield, Missouri

Awards

2009 Award winning paper (Association of Chiropractic Colleges Educational Conference / Research Agenda Conference)

Society Memberships

- 2010 – Present Sigma Xi
- 2010 – Present North American Spine Society
- 2009 – Present Biomedical Engineering Society
- 2007 – Present Orthopaedic Research Society
- 2007 – Present American Society of Biomechanics
- 2007 – Present International Society of Biomechanics
- 2001 – Present American Society of Mechanical Engineers

Research Support

- Scoliosis Research Society – New Investigator Grant (2007) $24,999

Title Biomechanics of cervical spine in patients with thoracic scoliosis
PI Mozammil Hussain, Ph.D.
Co-I Rodger Tepe, Ph.D., Norman W. Kettner, D.C. – Division of Research, Logan University, Chesterfield, Missouri
Consultant Lawrence G. Lenke, M.D. – Department of Orthopaedic Surgery, Washington University Medical Center, St. Louis, Missouri
Status score – 3, not funded

Peer-Reviewer for Journals

- Clinical Biomechanics
- Annals of Biomedical Engineering
- The Open Biomedical Engineering Journal
- Journal of Mechanics in Medicine and Biology

Peer-Reviewed Journal Papers

- Hussain M, Natarajan RN, Chaudhary G, An HS, Andersson GB. Relative contributions of strain-dependent permeability and fixed charged density of proteoglycans in predicting cervical disc biomechanics: a poroelastic C5-C6 finite element model study. *Medical Engineering & Physics.* (In Press).
- Hussain M, Natarajan RN, An HS, Andersson GB. Simulation of inhomogeneous rather than homogeneous poroelastic tissue material properties within disc annulus and nucleus better predicts cervical spine response: a C3-T1 finite element model analysis under compression and moment loadings. *Spine.* (In Press).
- Hussain M, Natarajan RN, An HS, Andersson GB. Reduction in segmental flexibility because of disc degeneration is accompanied with higher changes in facet loads than changes in disc pressure: a C5-C6 poroelastic finite element investigation. *The Spine Journal.* 10(12):1069-77, 2010.
- Hussain M, Natarajan RN, An HS, Andersson GB. Patterns of height changes in anterior and posterior cervical disc regions affects the contact loading at posterior facets during moderate and severe disc degeneration: a C5-C6 poroelastic finite element model study. *Spine.* 35(18):E873-E881, 2010.
- Hussain M, Gay RE, An KN. Reduction in disc and fiber stresses by axial distraction is higher in the disc with fibers oriented towards the vertical than horizontal plane: a finite element investigation. *Journal Of Manipulative And Physiological Therapeutics.* 33(4):252-260, 2010.
- Hussain M, Natarajan RN, An HS, Andersson GB. Motion changes in adjacent segments due to moderate and severe degeneration in C5-C6 disc – a poroelastic C3-T1 finite element model study. *Spine.* 35(9):939-947, 2010.
- Hussain M, Natarajan RN, Braaksma BR, Fayyazi AH, Andersson GB, An HS. Screw angulation affects bone-screw stresses and bone graft load-sharing in anterior cervical corpectomy fusion with a rigid screw-plate construct: a finite element model study. *The Spine Journal.* 9(12):1016-1023, 2009.

Conference & Symposium Proceedings

YEAR – 2011

- **Hussain M.** A biomechanical finite element study of the disc tissues with incompleteness and laxity in annular fibers. Proceedings of the Association of Chiropractic Colleges Educational Conference XVIII/Research Agenda Conference XVI, March 17-19, 2011, Las Vegas, Nevada. (Accepted).

YEAR – 2010

- Hussain, M, Gay RE, An KN. Inner trabaecular bone is more affected than outer cortical bone due to incomplete annular fibers. Fall Annual Meeting of Biomedical Engineering Society, October 6-9, 2010, Austin, Texas.
- Hussain, M, Gay RE, An KN. Cervical endplates are affected by incomplete length of collagen fibers: a finite element analysis. Fall Annual Meeting of Biomedical Engineering Society, October 6-9, 2010, Austin, Texas.
- Hussain, M, Gay RE, An KN. Incomplete fibers in inner annulus affects tissue stresses higher than in outer or middle annulus. Fall Annual Meeting of Biomedical Engineering Society, October 6-9, 2010, Austin, Texas.
- Hussain, M, Gay RE, An KN. Disc with incomplete fibers bulges more than the disc with complete fibers: a finite element study. Fall Annual Meeting of Biomedical Engineering Society, October 6-9, 2010, Austin, Texas.
- Hussain, M, Gay RE, An KN. Load bearing capability of annular fibers is affected by their incomplete length: a finite element model analysis. 34th Annual Meeting of the American Society of Biomechanics, August 18-21, 2010, Providence, Rhode Island.
- Hussain, M, Gay RE, An KN. Disc biomechanical response due to incomplete length of collagen fibers in various regions of annulus fibrosus: a finite element investigation. Summer Bioengineering Conference, American Society of Mechanical Engineers, June 16-19, 2010, Naples, Florida.
- Hussain M. A finite element analysis of the intervertebral disc with complete and incomplete length of collagen fibers: application of distraction forces on compressed discs. Proceedings of the Association of Chiropractic Colleges Educational Conference XVII/Research Agenda Conference XV, March 18-20, 2010, Las Vegas, Nevada. (Award Nomination Paper).

YEAR – 2009

- Hussain M, Gay RE, An KN, Triano JJ. Disc bulging decreases with degeneration in tissue matrix and increases with laxity in annular fibers. Fall Annual Meeting of Biomedical Engineering Society, October 7-10, 2009, Pittsburgh, Pennsylvania.
- Hussain M, Gay RE, An KN, Triano JJ. Increase in outer annulus and decrease in annulus-nucleus bulgings during cervical disc degeneration. Fall Annual Meeting of Biomedical Engineering Society, October 7-10, 2009, Pittsburgh, Pennsylvania.
- Hussain M, Gay RE, An KN, Triano JJ. Annular fibers in disc matrix alter hard and soft tissues in compression: a finite element analysis. Fall Annual Meeting of Biomedical Engineering Society, October 7-10, 2009, Pittsburgh, Pennsylvania.

- Hussain M, Gay RE, An KN, Triano JJ, Tepe R. Relative contributions to disc degeneration progression is higher by degenerative tissue ground matrix than annular fibers laxity: a finite element analysis in compression. 33rd Annual Meeting of the American Society of Biomechanics, August 26-29, 2009, Pennsylvania State University, Pennsylvania.
- Hussain M, Gay RE, An KN, Tepe R. Biomechanical effects of chiropractic-based distraction therapy on the behavior of healthy and degenerated cervical discs: a finite element model analysis. Summer Bioengineering Conference, American Society of Mechanical Engineers, June 17-21, 2009, Lake Tahoe, California.
- Hussain M, Tepe R. Biomechanical role of annular fiber orientation angle under compression and traction loadings. Proceedings of the Association of Chiropractic Colleges Educational Conference XVI/Research Agenda Conference XIV, March 12-14, 2009, Las Vegas, Nevada. (Award Winning Paper).
- Hussain M, Natarajan RN, An HS, Andersson GBJ. Progressive disc degeneration at any intermediate segment affects disc and posterior facets mechanics of inferior segment more than superior segment – a finite element model study. 55th Annual Meeting of Orthopaedic Research Society, February 22-25, 2009, Las Vegas, Nevada.
- Hussain M, Natarajan RN, An HS, Andersson GBJ. Segmental mechanics between anterior disc, posterior disc, and posterior facets at moderate and severe stages of cervical disc degeneration – a finite element model study. 55th Annual Meeting of Orthopaedic Research Society, February 22-25, 2009, Las Vegas, Nevada.

YEAR – 2008

- Hussain M, Tepe R. Tissue damage to an injured disc and position of nucleus pulposus – a finite element investigation. Fall Annual Meeting of Biomedical Engineering Society, October 2-4, 2008, St. Louis, Missouri.
- Hussain M, Tepe R. Stages of disc degeneration and biomechanics of vertebra, endplate and disc – a finite element study. Fall Annual Meeting of Biomedical Engineering Society, October 2-4, 2008, St. Louis, Missouri.
- Hussain M, Tepe R. Disc height reduction is a better predictor of cervical disc degeneration progression than reduction in the area of nucleus pulposus – a finite element analysis. 4th North American Congress on Biomechanics, August 5-9, 2008, Ann Arbor, Michigan.
- Hussain M, Tepe R. Effect of traction load on the biomechanical behavior of cervical, thoracic, and lumbar spinal segments – a finite element analysis. Summer Bioengineering Conference, American Society of Mechanical Engineers, June 25-29, 2008, Marco Island, Florida.
- Hussain M, Tepe R. Peripheral, circumferential, and radial components of stresses and strains within different regions of annulus and nucleus – a cervical intervertebral disc under pure compression. Proceedings of the Association of Chiropractic Colleges Educational Conference XV/Research Agenda Conference XIII, March 13-15, 2008, Washington DC.
- Hussain M, Tepe R. High stresses in the anterior-lateral region of annulus due to strains in the nucleus – a cervical intervertebral disc under pure compression. Proceedings of the Association of Chiropractic Colleges Educational Conference XV/Research Agenda Conference XIII, March 13-15, 2008, Washington DC.

YEAR – 2007

- Bresnahan LE, Hussain M, Natarajan RN, Andersson GBJ, Fessler RG. Biomechanical effects of minimally invasive decompression of cervical stenosis (MEDS): a finite element investigation. 57th Annual Meeting of Congress of Neurological Surgeons, September 15-20, 2007, San Diego, California.
- Bresnahan LE, Hussain M, Fessler RG, Natarajan RN, Andersson GBJ. A finite element investigation into the biomechanical effects of minimally invasive treatment for cervical spondylotic myleopathy. 31st Annual Meeting of the American Society of Biomechanics, August 22-25, 2007, Stanford University, California.
- Hussain M, Natarajan RN, Andersson GBJ, An HS. Stresses on movable core and loads on facets are higher by implanting cervical artificial disc prosthesis as compared to bone grafting fusion technique: a finite element model study. 31st Annual Meeting of the American Society of Biomechanics, August 22-25, 2007, Stanford University, California.
- Hussain M, Nassr AN, Natarajan RN, Andersson GBJ, An HS. Biomechanics of adjacent segments with number of inter-body grafts and spinal instrumentations for a multi-level fusion construct using a finite element model. 31st Annual Meeting of the American Society of Biomechanics, August 22-25, 2007, Stanford University, California.
- Hussain M, Nassr AN, Natarajan RN, Andersson GBJ, An HS. Effect of corpectomy and discectomy fusion procedures on the stability of multi-level cervical construct with anterior rigid screw-plate fixation – a finite element model study. Summer Bioengineering Conference, American Society of Mechanical Engineers, June 20-24, 2007, Keystone, Colorado.
- Hussain M, Natarajan RN, Andersson GBJ, An HS. Effect of a degenerated C5-C6 disc on the biomechanics of adjacent levels – a poroelastic finite element investigation. Summer Bioengineering Conference, American Society of Mechanical Engineers, June 20-24, 2007, Keystone, Colorado.
- Hussain M, Natarajan RN, Andersson GBJ, An HS. Biomechanical response of a poro-elastic cervical spine under compression. 53rd Annual Meeting of the Orthopaedic Research Society, February 11-14, 2007, San Diego, California.
- Hussain M, Natarajan RN, Andersson GBJ, An HS. Investigation on the effect of degenerative changes in the biomechanics of a C5-C6 motion segment – a poroelastic finite element model study. 15th Annual Symposium of Computational Methods in Orthopaedic Biomechanics, February 10, 2007, San Diego, California.

YEAR – 2006

- Hussain M, Nassr AN, Natarajan RN, Andersson GBJ, An HS. Biomechanical stabilization of the multi-level corpectomy construct using anterior and posterior instrumentation – a finite element model study. 30th Annual Meeting of American Society of Biomechanics, September 6-9, 2006, Blacksburg, Virginia.
- Hussain M, Fayyazi AH, Natarajan RN, Andersson GBJ, An HS. Effect of uni- and bi-cortical screw angulation on the stability of multi-level corpectomy construct – a finite element model study. 30th Annual Meeting of American Society of Biomechanics, September 6-9, 2006, Blacksburg, Virginia.
- Hussain M, Natarajan RN, Andersson GBJ, An HS. Changes in biomechanical response due to disc degeneration: a poroelastic C5-C6 motion segment study. Summer Bioengineering Conference, American Society of Mechanical Engineers, June 21-25, 2006, Amelia Island, Florida.

- Natarajan RN, Hussain M, An HS, Andersson GBJ. Change in the biomechanics of lumbar spine due to artificial disc placement. 33rd Annual Meeting of The International Society for the Study of the Lumbar Spine, June 13-17, 2006, Bergen, Norway.
- Natarajan RN, Hussain M, Manson N, Reo M, Robinson J, An HS, Andersson GBJ. Biomechanical comparison of the intact cervical disc versus total disc arthroplasty: a finite element model study. 6th Annual Meeting of Spine Arthroplasty Society, May 9-13, 2006, Montreal, Canada.
- Nassr AN, Hussain M, Natarajan RN, An HS, Andersson GB, Singh K. Long-segment spinal reconstruction in the setting of multilevel cervical corpectomies: a finite element analysis. 9th Annual Meeting of Midwest Resident Research Symposium, April 27, 2006, Chicago, Illinois.
- Natarajan RN, Hussain M, Andersson GBJ, An HS. Strain-dependent permeability has a predominant effect over swelling pressure in predicting biomechanics of cervical spine. 52nd Annual Meeting of the Orthopaedic Research Society, March 19-22, 2006, Chicago, Illinois.
- Natarajan RN, Hussain M, Andersson GBJ, An HS. Development of a physiologically relevant numerical model to study biomechanics of degenerated cervical spine. 52nd Annual Meeting of the Orthopaedic Research Society, March 19-22, 2006, Chicago, Illinois.
- Natarajan RN, Hussain M, An HS, Andersson GBJ. The effect of placement of an artificial disc on the biomechanics of a lumbar motion segment – a finite element model study. 52nd Annual Meeting of the Orthopaedic Research Society, March 19-22, 2006, Chicago, Illinois.
- Hussain M, Natarajan RN, Andersson GBJ, An HS. Comparison of the biomechanical response between non-linear elastic and poroelastic finite element models of a healthy C5-C6 motion segment. 14th Annual Symposium of Computational Methods in Orthopaedic Biomechanics, March 18, 2006, Chicago, Illinois.

YEAR – 2005

- Hussain M, Natarajan RN, Andersson GBJ, An HS. Two-level cervical corpectomy with rigid screw-plate system produces larger increase in facet loads at level superior to the fusion level. Summer Bioengineering Conference, American Society of Mechanical Engineers, June 22-26, 2005, Vail, Colorado.
- Fayyazi AH, Hussain M, Natarajan RN, Phillips FM, An HS. Effect of graft-endplate contact surface morphology on the biomechanics of anterior cervical corpectomy fusion and rigid plating: a finite-element model study. 3rd Annual Meeting of World Spine Society, August 1-3, 2005, Rio de Janeiro, Brazil.
- Fayyazi AH, Hussain M, Natarajan RN, Phillips FM, An HS. Effect of screw angulation on the bone graft load sharing and bone-screw stresses in anterior cervical corpectomy fusion with rigid plate screw construct: a finite-element model study. 3rd Annual Meeting of World Spine Society, August 1-3, 2005, Rio de Janeiro, Brazil.
- Fayyazi AH, Hussain M, Natarajan RN, Phillips FM, An HS. Effect of screw angulation on the bone graft load sharing and bone-screw stresses in anterior cervical corpectomy fusion with rigid plate screw construct: a finite-element model study. 12th International Meeting on Advanced Spine Techniques, July 7-9, 2005, Alberta, Canada.

YEAR – 2004

- Fayyazi AH, Hussain M, Natarajan RN, Phillips FM, An HS. Effect of screw angulation on the bone graft load sharing and bone-screw stresses in anterior cervical corpectomy fusion with rigid plate screw construct: a finite-element model study. 2nd Annual Depuy Spine Fellows Course, June, 2004, Boston, Massachusetts (Best Basic Science Paper).
- Hussain M, Natarajan RN, Deol GS, Andersson GBJ, An HS. Stability of corpectomy with anterior cervical plating depends on the angular placement of screws: a FE model study. ASME International Mechanical Engineering Congress and RD & D Expo, November 13-19, 2004, Anaheim, California.
- Natarajan RN, Hussain M, Wimmer MA, Rosenberg AG, Jacobs JJ. Wear of UHMWPE components of total knee replacements depend upon walking pattern. 50th Annual Meeting of the Orthopaedic Research Society, March 7-10, 2004, San Francisco, California.

Invited Presentations

- Hussain M. Understanding degeneration progression along the cervical spine using finite element computer modeling technique. Department of Physical Therapy, The City University of New York, Staten Island, New York, April 14, 2010.
- Hussain M. Biomechanical effects of spinal implants on the behavior of surgically fused cervical motion segments. Department of Physical Therapy, The City University of New York, Staten Island, New York, April 14, 2010.
- Hussain M. Effect of anterior and/or posterior instrumentation on the stability of multi-level cervical spine fusion constructs – a finite element study. Division of Research, Logan University, Chesterfield, Missouri, January 30, 2007.
- Hussain M. Progression of disc degeneration along the cervical spine – a poroelastic finite element study. Department of Orthopaedic Surgery, Beth Israel Deaconess Medical Center, Harvard Medical School, Boston, Massachusetts, January 22, 2007.
- Hussain M. Propagation of disc degeneration along the cervical spine – a poroelastic finite element investigation. Division of Research, Logan University, Chesterfield, Missouri, December 11, 2006.
- Hussain M. Biomechanical response due to disc degeneration of a poroelastic C5-C6 motion segment. Division of Research, Logan University, Chesterfield, Missouri, December 11, 2006.

In: Anatomy and Physiology Research
Editors: N. R. Zirena and X. B. Liu
ISBN: 978-1-62942-595-8

Chapter 11

Biographical Sketch - Anatomy and Physiology Research

Makoto Ishikawa

Affiliation:

M.D. & Ph.D.
Department of Ophthalmology Akita Graduate University School of Medicine

Address:

Department of Ophthalmology Akita Graduate University School of Medicine
Hondo 1-1-1, Akita 010-8543, Japan

Date of Birth:

July 18. 1959

Education:

1979-1985: M.D., Yamagata University School of Medicine
1985-1989: M.D. & Ph.D.;
Anatomy and Otolaryngology,
Yamagata University School of Medicine

Professional Appointments:

1989-1992 Instructor, Department of Anatomy, Yamagata University School of Medicine

1991-1992 Postdoctoral fellow, Department of Anatomy and Cell Biology, University of Southern California School of Medicine, (sponsored by Japanese Ministry of Education, Science, and Culture)

1993-1994 Senior Resident, Department of Ophthalmology Akita University School of Medicine, 1994-2000 Instructor, Department of Ophthalmology Akita University School of Medicine

1996-Board certified specialist in Japanese Ophthalmological Society

2001-2002 Assistant Professor, Department of Ophthalmology, Akita University School of Medicine

2002-2005 Chairperson,Department of Ophthalmology, Ogachi Central Hospital (Akita)

2006- Associate Professor, Department of Ophthalmology Akita University School of Medicine

Publications Last 3 Years:

1. Ishikawa M, Yoshitomi T, Zorumski CF, Izumi Y.Effects of acutely elevated hydrostatic pressure in the rat ex vivo retinal preparation. Invest Ophthalmol Vis Sci Publish ahead of print for 4 Aug 2010
2. Sawada Y, Ishikawa M, Yoshitomi T.Optic nerve head morphology assessed by laser scanning tomography in normal Japanese subjects. J Glaucoma 2010. 7.22 accept (now in printing)
3. Ishikawa M, Fujiwara T, Yoshitomi T.Temperature-dependent ultrastructural changes in the cone interphotoreceptor matrix.Jpn J Ophthalmol 2009;53:536-540.
4. Ishikawa M, Tatabe H, Hayakawa M, Yoshitomi T.Peripheral T-cell Lymphoma of the Eyelid. Clinical Ophthalmol 2009;3:527-529
5. Ishikawa M, Yoshitomi T. Effects of brinzolamide vs. timolol as an adjunctive medication to latanoprost on circadian intraocular pressure control in primary open-angle glaucoma Japanese patients. Clinical Ophthalmol 2009;3:493-500.
6. Inoue H, Sonoda K, Ishikawa M, et al. Clinical evaluation of local ocular toxicity in candidate anti-adenoviral agents in vivo. Ophthalmologica 2009;223:233-238.
7. Ishikawa M, Yoshitomi T. Effects of brinzolamide vs. timolol as an adjunctive medication to latanoprost on circadian intraocular pressure control in primary open-angle glaucoma Japanese patients. Clinical Ophthalmol 2009;30:493-500.
8. Inoue H, Sonoda K, Ishikawa M, Kadonosono K, Uchio E. Clinical evaluation of local ocular toxicity in candidate anti-adenoviral agents in vivo.Ophthalmologica 2009;223:233-238.

In: Anatomy and Physiology Research
Editors: N. R. Zirena and X. B. Liu
ISBN: 978-1-62942-595-8

Chapter 12

Biographical Sketch - Anatomy and Physiology Research

Yuichi Ishikawa

Affiliation:

The Cancer Institute, Japanese Foundation for Cancer Research

Address:

3-8-31 Ariake, Koto-ku, Tokyo 135-8550

Date of Birth:

May 9, 1953

Education:

1977 - Graduated from University of Tokyo, Faculty of Science; Geophysics
1979 - M. D.; Finished graduate course of University of Tokyo; Geophysics
1985 - Graduated from Tokyo Medical and Dental University Medical School
1989 - Finished graduate course of Tokyo Medical and Dental University; Pathology
1989 - Ph. D.; Radiation carcinogenesis

Research and Professional Experience:

1986; Resident, Pathology Department of Japanese Red Cross Medical Center, Tokyo

1986 1987; Resident, Pathology Department of JFCR Cancer Institute, Tokyo
1987; Student, Radiation Protection Course of National Institute of Radiological Sciences, Chiba
1987 1988; Staff Pathologist, Pathology Department of Tokyo Metropolitan Komagome Hospital, Tokyo
1989 1990; Research Associate, Department of Pathology, JFCR Cancer Institute, Tokyo
1990 1996; Associate, Department of Pathology, JFCR Cancer Institute, Tokyo
1992 1993; Visiting Scientist, Biomedical Department, Harwell Laboratory, United Kingdom Atomic Energy Authority, Oxford, UK
1996 2003; Associate Member, Department of Pathology, JFCR Cancer Institute, Tokyo
2003 2004: Vice Chief, Department of Pathology, JFCR Cancer Institute, Tokyo
2005 present: Member and Chief, Division of Pathology, JFCR Cancer Institute, Tokyo

Professional Appointments:

Member and Chief, Division of Pathology, The Cancer Institute,
Japanese Foundation for Cancer Research

Publications Last 3 Years:

2011

Sakao Y, Okumura S, Mun M-y, Uehara H, Ishikawa Y, Nakagawa K. The impact of superior mediastinal lymph node metastases on prognosis in non-small cell lung cancer located in the right middle lobe. J Thorac Oncol 2011 Mar;6(3):494-9.

2010

Migita T, Narita T, Asaka R, Miyagi E, Nagano H, Nomura K, Matsuura M, Satoh Y, Okumura S, Nakagawa K, Seimiya H, Ishikawa Y. Role of insulin-like growth factor binding proteins 2 in lung cancer: IGF independent anti-apoptotic effect via caspase-3. Am J Pathol, 176: 1756-66, 2010.

Inamura K, Ishikawa Y. Most Common Cancers in Asia-Pacific Region/Lung Cancer. Asian Pacific J Cancer Prev, in press.

Hiramatsu M, Ninomiya H, Inamura K, Nomura K, Takeuchi K, Satoh Y, Okumura S, Nakagawa K, Yamori T, Matsuura M, Morikawa T, Ishikawa Y. Activation status of receptor tyrosine kinase downstream pathways in primary lung adenocarcinoma with reference of KRAS and EGFR mutations. Lung Cancer 2010, 70: 94-102.

Inamura K, Ninomiya H, Ishikawa Y, Matsubara O. Is the epidermal growth factor receptor status in lung cancers reflected in clinicopathologic features? Arch Pathol Lab Med. 2010 Jan;134(1):66-72.

Satoh Y, Ishikawa Y. Primary pulmonary meningioma: Ten-year follow-up findings for a multiple case, implying a benign biological nature. J Thorac Cardiovasc Surg. 2010 Mar;139(3): e39-40.

Hoshi R, Furuta N, Horai T, Ishikawa Y, Miyata S, Satoh Y. Discriminant model for cytologic distinction of large cell neuroendocrine carcinoma from small cell carcinoma of the lung. J Thorac Oncol. 2010 (4); 5: 472-8.

Fujino T, Nomura K, Ishikawa Y, Makino H, Umezawa A, Aburatani H, Nagasaki K, Nakamura T. Function of EWS-POU5F1 in sarcomagenesis and tumor cell maintenance. Am J Pathol. 2010 Apr;176(4):1973-82.

Inamura K, Ishikawa Y. Lung cancer progression and metastasis from the prognostic point of view. Clin Exp Metastasis. (2010) 27:389-97

Sakao Y, Okumura S, Mun M, Uehara H, Ishikawa Y, Nakagawa K. Prognostic heterogeneity in multilevel N2 non-small cell lung cancer patients: importance of lymphadenopathy and occult intrapulmonary metastases. Ann Thorac Surg. 2010 Apr;89(4):1060-3.

Okada A, Shimmyo T, Hashimoto T, Kobayashi Y, Miyagi Y, Ishikawa Y, Nakagawa K, Hayashi J, Tsuchiya E. Predictive advantage of a cell type classification for pulmonary adenocarcinoma coupled with data for p53, K-ras and EGFR alterations. Cancer Sci. 2010 Jul;101(7):1745-53. Epub 2010 Apr 7.

Mochizuki T, Okumura S, Ishii G, Ishikawa Y, Hayashi R, Kawabata K, Yoshida J. Surgical resection for oral tongue cancer pulmonary metastases. Interact Cardiovasc Thorac Surg. 2010 Jul;11(1):56-9. Epub 2010 Mar 31.

Yanai H, Nakamura K, Hijioka S, Kamei A, Ikari T, Ishikawa Y, Shinozaki E, Mizunuma N, Hatake K, Miyajima A. Dlk-1, a cell surface antigen on foetal hepatic stem/progenitor cells, is expressed in hepatocellular, colon, pancreas and breast carcinomas at a high frequency. J Biochem. 2010 Jul;148(1):85-92. Epub 2010 Mar 30.

Ohba T, Motoi N, Kimura Y, Okumura S, Kawabata K, Yoshizawa Y, Inase N, Ishikawa Y. Cytokeratin expression profiling is useful for distinguishing between primary squamous cell carcinoma of the lung and pulmonary metastases from tongue cancer. Pathol Int. 2010 Aug;60(8):575-80.

Dan S, Okamura M, Seki M, Yamazaki K, Sugita H, Okui M, Mukai Y, Nishimura H, Asaka R, Nomura K, Ishikawa Y, Yamori T. Correlating phosphatidylinositol 3-kinase inhibitor efficacy with signaling pathway status: in silico and biological evaluations. Cancer Res. 2010 Jun 15;70(12):4982-94. Epub 2010 Jun 8.

Yamamoto S, Kawakami S, Yonese J, Fujii Y, Tsukamoto T, Okubo Y, Kijima T, Ishikawa Y, Fukui I. Feasibility of antegrade radical prostatectomy for clinically locally advanced prostate cancer: a comparative study with clinically localized disease. Int J Urol. 2010, 17(8):720-5.

Uehara H, Sakao Y, Mun M, Nakagawa K, Nishio M, Ishikawa Y, Okumura S. Prognostic value and significance of subcarinal and superior mediastinal lymph node metastasis in lower lobe tumours. Eur J Cardiothorac Surg. 38 (2010) 498-502.

Okamoto S, Machinami R, Tanizawa T, Matsumoto S, Lee GH, Ishikawa Y. Dedifferentiated liposarcoma with rhabdomyoblastic differentiation in an 8-year-old girl. Pathol Res Pract. 2010;206(3):191-196.

Takeuchi S, Takahashi A, Motoi N, Yoshimoto S, Tajima T, Yamakoshi K, Hirao A, Yanagi S, Fukami K, Ishikawa Y, Sone S, Hara E, Ohtani N. Intrinsic cooperation between p16INK4a and p21Waf1/Cip1 in the onset of cellular senescence and tumor suppression in vivo. Cancer Res. 2010 Nov 15;70(22):9381-90. Epub 2010 Nov 9.

2009

Komai Y, Fujiwara M, Fujii Y, Mukai H, Yonese J, Kawakami S, Yamamoto S, Migita T, Ishikawa Y, Kurata M, Nakamura T, Fukui I. Adult Xp11 translocation renal cell carcinoma diagnosed by cytogenetics and immunohistochemistry. Clin Cancer Res, 15(4):1170-6, 2009.

Ninomiya H, Hiramatsu M, Inamura K, Nomura K, Okui M, Miyoshi T, Okumura S, Satoh Y, Nakagawa K, Nishio M, Horai T, Miyata S, Tsuchiya E, Fukayama M, Ishikawa Y. Correlation between morphology and EGFR mutations in lung adenocarcinomas. Significance of the micropapillary pattern and the hobnail cell type. Lung Cancer 63(2):235-40, 2009.

Takeuchi K, Choi YL, Togashi Y, Soda M, Hatano S, Inamura K, Takada S, Ueno T, Yamashita Y, Satoh Y, Okumura S, Nakagawa K, Ishikawa Y, Mano H. KIF5B-ALK, a novel fusion oncokinase identified by an immunohistochemistry-based diagnostic system for ALK-positive lung cancer. Clin Cancer Res, 15(9):3143-9, 2009.

Inamura K, Takeuchi K, Togashi Y, Hatano S, Ninomiya H, Motoi N, Mun MY, Sakao Y, Okumura S, Nakagawa K, Soda M, Lim Choi Y, Mano H, Ishikawa Y. EML4-ALK lung cancers are characterized by rare other mutations, a TTF-1 cell lineage, an acinar histology, and young onset. Mod Pathol. 22: 508-515, 2009.

Kato M, Sanada M, Kato I, Sato Y, Takita J, Takeuchi K, Niwa A, Chen Y, Nakazaki K, Nomoto J, Asakura Y, Muto S, Tamura A, Iio M, Akatsuka Y, Hayashi Y, Mori H, Igarashi T, Kurokawa M, Chiba S, Mori S, Ishikawa Y, Okamoto K, Tobinai K, Nakagama H, Nakahata T, Yoshino T, Kobayashi Y, Ogawa S. Frequent inactivation of A20 in B-cell lymphomas. Nature 459: 712-6, 2009.

Yamamoto S, Kawakami S, Yonese J, Fujii Y, Tsukamoto T, Ohkubo Y, Komai Y, Ishikawa Y, Fukui I. Risk stratification of high-grade prostate cancer treated with antegrade radical prostatectomy with intended wide resection. Jpn J Clin Oncol 39(6):387-93, 2009.

Nakamura Y, Migita M, Hosoda F, Okada N, Gotoh M, Arai Y, Fukushima M, Ohki M, Miyata S, Takeuchi K, Imoto I, Katai H, Yamaguchi T, Inazawa J, Hirohashi S, Ishikawa Y, Shibata T. Krüppel-like factor 12 plays a significant role in poorly differentiated gastric cancer progression. Int J Cancer 125:1859-67, 2009.

Makino H, Toyoda M, Matsumoto K, Saito H, Nishino K, Fukawatase Y, Machida M, Akutsu H, Uyama T, Miyagawa Y, Okita H, Kiyokawa N, Fujino T, Ishikawa Y, Nakamura T, Umezawa A. Mesenchymal to embryonic incomplete transition of human cells by chimeric OCT4/3 (POU5F1) with physiological co-activator EWS. Exp Cell Res. 315(16): 2727-40, 2009.

Satoh Y, Hoshi R, Horai T, Okumura S, Nakagawa K, Ishikawa Y, Miyata S. Association of cytologic micropapillary clusters in cytology samples with lymphatic spread in clinical stage I lung adenocarcinomas. Lung Cancer 2009 Jun;64(3):277-81.

Hanyu A, Kojima K, Hatake K, Nomura K, Murayama H, Ishikawa Y, Miyata S, Ushijima M, Matsuura M, Ogata E, Miyazawa K, Imamura T. Functional in vivo optical imaging of tumor angiogenesis, growth, and metastasis prevented by administration of anti-human VEGF antibody in xenograft model of human fibrosarcoma HT1080 cells. Cancer Sci. 2009;100:2085-92.

2008

Inamura K, Takeuchi K, Togashi Y, Nomura K, Ninomiya H, Okui M, Satoh Y, Okumura S, Nakagawa K, Soda M, Choi YL, Niki T, Mano H, Ishikawa Y. EML4-ALK fusion is linked to histological characteristics in a subset of lung cancers. J Thorac Oncol, 3: 13-7, 2008.

Shimmyo T, Okada A, Hashimoto T, Kobayashi Y, Miyagi Y, Ishikawa Y, Nakagawa K, Osada H, Tsuchiya E.Etiologic value of p53 mutation spectra and differences with histology in lung cancers. Cancer Sci. 99: 287-95, 2008.

Ohyanagi F, Horiike A, Okano Y, Satoh Y, Okumura S, Ishikawa Y, Nakagawa K, Horai T, Nishio M. Phase II trial of gemcitabine and irinotecan in previously treated patients with small-cell lung cancer. Cancer Chemother Pharmacol. 61: 503-8, 2008.

Futami K, Ishikawa Y, Goto M, Furuichi Y, Sugimoto M. Role of Werner syndrome gene product helicase in carcinogenesis and in resistance to genotoxins by cancer cells. Cancer Sci 99: 843-8, 2008.

Tsuzuki T, Ninomiya H, Natori Y, Ishikawa Y. Coalescent pleural malignant mesothelioma and adenocarcinoma of the lung, involving only minor asbestos exposure. Pathol Int 58: 451-5, 2008.

Choi YL, Takeuchi K, Soda M, Inamura K, Togashi Y, Hatano S, Enomoto M, Hamada T, Haruta H, Watanabe H, Kurashina K, Hatanaka H, Ueno, T, Takada S, Yamashita Y, Sugiyama Y, Ishikawa Y, Mano H. Identification of novel isoforms of the EML4-ALK transforming gene in nonsmall cell lung cancer. Cancer Res, 68: 4971-6, 2008.

Inamura K, Togashi Y, Ninomiya H, Shimoji T, Noda T, Ishikawa Y. HOXB2, an adverse prognostic indicator for stage I lung adenocarcinomas, promotes invasion by transcriptional regulation of metastasis-related genes in HOP-62 non-small cell lung cancer cells. Anticancer Res, 28: 2121-8, 2008.

Takeuchi K, Choi YL, Soda M, Inamura K, Togashi Y, Hatano S, Enomoto M, Takada S, Yamashtia Y, Satoh Y, Okumura S, Nakagawa K, Ishikawa Y, Mano H. Multiplex RT-PCR screening for EML4-ALK fusion transcripts. Clin Cancer Res 14: 6618-24, 2008.

Migita T, Narita T, Nomura K, Miyagi E, Inazuka F, Matsuura M, Ushijima M, Mashima T, Seimiya H, Satoh Y, Okumura S, Nakagawa K, Ishikawa Y. ATP Citrate Lyase: Activation and Therapeutic Implications in Non-small Cell Lung Cancer. Cancer Res 68: 8547-54, 2008.

Katsuno Y, Hanyu A, Kanda H, Ishikawa Y, Akiyama F, Iwase T, Ogata E, Ehata S, Miyazono K, Imamura T. Bone morphogenetic protein signaling enhances invasion and bone metastasis of breast cancer cells through Smad pathway. Oncogene 27: 6322-33, 2008.

Soda M, Takada S, Takeuchi K, Choi Y-L, Enomoto M, Ueno T, Haruta H, Hamada T, Yamashita Y, Ishikawa Y, Sugiyama Y, Mano H. A mouse model for EML4-ALK-positive lung cancer. Proc Natl Acad Sci USA 105: 19893-7, 2008.

Tatokoro M, Kawakami S, Yonese J, Fujii Y, Okubo Y, Yamamoto S, Takeshita H, Komai Y, Ishikawa Y, Fukui I. Preliminary report of multimodal treatment with ifosfamide, 5-fluorouracil, etoposide and cisplatin (IFEP chemotherapy) against metastatic adenocarcinoma of the urachus. Int J Urol. 2008 Sep;15(9):851-3.

Machinami R, Nishida K, Ishida T, Matsumoto S, Kuroda K, Kobayashi M, Takeuchi K, Ishikawa Y. Carcinosarcomatous malignancy, osteosarcoma and squamous cell carcinoma, in giant cell tumor of the right distal femur. Pathol Res Pract. 2008;204(8):583-8.

Hiramatsu M, Inagaki T, Inagaki T, Matsui Y, Satoh Y, Okumura S, Ishikawa Y, Miyaoka E, Nakagawa K. Pulmonary ground-glass opacity (GGO) lesions-large size and a history of lung cancer are risk factors for growth. J Thorac Oncol. 3(11):1245-50, 2008.

Sugai S, Satoh Y, Komatsu M, Okumura S, Nakagawa K, Ishikawa Y, Miki Y. Recurrence pattern and rapid intraoperative detection of carcinoembryonic antigen (CEA) mRNA in pleural lavage in patients with non-small cell lung cancer (NSCLC). Rinsho Byori 56:851-1, 2008.

In: Anatomy and Physiology Research
Editors: N. R. Zirena and X. B. Liu
ISBN: 978-1-62942-595-8

Chapter 13

Biographical Sketch - Anatomy and Physiology Research

Nikolai I. Kononenko

Affiliation:

Dr.

Address:

E-mail: N.I.Kononenko@gmail.com; N.I.Kononenko@biph.kiev.ua

Date of Birth:

December 15, 1946

Education:

1970 - Donetsk State University, Faculty of Physics (Donetsk, Ukraine), Postgrad. Student
1972-1975 - Bogomoletz Institute of Physiology (Kiev, Ukraine)
1976 - Ph.D.; Thesis supervisor Prof. Platon G. Kostyuk

Research and Professional Experience:

Neurobiology, Circadian rhythms, Spontaneous activity, Plasticity, Suprachiasmatic neurons, Hippocampal neurons, Whole cell and On-cell recordings, Imaging, Calcium and GFP fluorimetry.

Professional Appointments:

Scientific researcher, Bogomoletz Institute of Physiology, Departmet of General Physiology of Nervous System.

Publications Last 3 Years:

Kononenko NI, Kuehl-Kovarik MC, Partin KM, Dudek FE, 2008. Circadian difference in firing rate of isolated rat suprachiasmatic nucleus neurons. Neurosci. Lett. 436: 314-316.

Kononenko NI, Honma S, Dudek FE, Honma K, 2008. On the role of calcium and potassium currents in circadian modulation of firing rate in rat suprachiasmatic nucleus neurons: multielectrode array analysis. Neurosci. Res. 62: 51-57.

Kononenko NI, Berezetskaya NM, 2010. Modeling the spontaneous activity in suprachiasmatic nucleus neurons: role of cation single channels. J. Theor. Biol. 265: 115-125.

In: Anatomy and Physiology Research
Editors: N. R. Zirena and X. B. Liu
ISBN: 978-1-62942-595-8

Chapter 14

Biographical Sketch - Anatomy and Physiology Research

Stefan Kujanik Sr.

Address:

Fax +421-55-6423763, e-mail: stefan.kujanik@upjs.sk

Date of Birth:

08.05.1949

Education:

P.J.Safarik University, School of Medicine in Kosice, Slovakia

Research and Professional Experience:

The cardiovascular system, The respiratory system, Biological rhythms

Professional Appointments:

Become an employee at School of Medicine as Assistant of Professor (1973)

PhD study with the thesis "Investigation of cardiac arrhythmias during stimulation of the hypothalamic nuclei" in the Moscow 2nd Medical Academy with the defence of PhD thesis supevised by professor G.I.Kosicky, MD, DSc (1976-1980)

Main investigator of the State research scientific project "Interrelationships of respiratory pathology and cardiac arrhythmias" (1981-1985)

Short-time research study stay at the Experimental department of IKEM Prague (1985)

Pedagogical Course for university teachers with the project "Importance of the continuous control of the study in education" (1986)

Promoted to Associate Professor degree (1987)

Main investigator of the State research (scientific) project "Respiratory disorders and origin of cardiac dysrhythmias" (1986-1990)

Patent No 244971 from 30/08/1988 (in Slovakia) "S. Kujanik, H. Jalcova, P. Jalc: Device for closing and opening the arterial lumen"

Short-time research study stay at the Dept of Physiology Medizinische Akademie in Magdeburg (German Democratic Republic 1987 and 1989)

Head of the Department of Physiology, School of Medicine in Kosice, Slovakia (1992-1997)

Main investigator of the National grant project "Dependence of periodicity of cardiac arrhythmias in respiratory diseases on altitude" (1994-1996)

Short-time research study and university management stay at the University of Gent, Belgium (1995)

Short-time stay during TEMPUS University management program (1996) at the University of Gent (Belgium) and Charing Cross and Westminster Medical School London (United Kingdom)

1997 - short-time research and pedagogic study stay (Tempus S-JEP project 11304) at the University of Veterinary Medicine Vienna (Austria)

1998 - short-time research study stay at the University of Bristol (United Kingdom)

1998 - short-time research study stay (Tempus S-JEP project 11304) at the University of Gent (Belgium) and Charing Cross and Westminster Medical School, London (United Kingdom)

2008-2010 - the main investigator of project of the National grant agency VEGA "Mechanisms of severe cardiovascular complications origin investigated by non-invasive methods" the author of 3 monographs, 8 university textbooks, 4 invited lectures in other countries, about 250 published papers in scientific literature.

Honors:

Graduated at the School of Medicine, P.J.Safarik University in Kosice with red diploma and distinction of the Dean (1973)

Award of the Czechoslovak Physiological Society for the best student research (scientific) work in Physiology 1983 for two my students.

Award of the Slovak Society for Physiology and Pathology of Respiration for the best publication 1984.

Membership in the Slovak Committee for Physiology and Pathology of Respiration (1987-1990)

Award of the Slovak Society for Physiology and Pathology of Respiration for the best publication 1992.

Contact Person of the Tempus S-JEP project 11304: "Strategic University Management in Administration" - for the P.J.Safarik University in Kosice (1996-1998)

Member of the Programme Committee of the XXIVth International Congress on Electrocardiology, Bratislava, June 24-28, 1997

1997-2000 - Committee member of the Slovak Society for Physiology and Pathology of Breathing.

Promoted to a professor position in the branch of Normal and pathological physiology (2003)

Promoted by the President of the Slovak republic to the degree of full university professor in the branch of Normal and pathological physiology (2006)

Since 2008 - member of the Editorial Board of the journal "Slovak Physician"

Invited Lectures in Abroad:

Kujanik S: Electrophysiological aspects of cardio-respiratory relations. Local session of the British Physiological Society, Charing Cross and Westminster Medical School, London, United Kingdom, October 9, 1996.

Kujanik S: Electrocardiological aspects of changed pulmonary ventilation and the moderate altitude. Local session of the Belgian Physiological Society - Faculty of Medicine, University of Ghent, Belgium, October 22, 1996.

Kujanik S: System of Teaching Physiology at the Medical Faculty of Safarik University, Kosice. Local session of the British Physiological Society, Charing Cross and Westminster Medical School, London, United Kingdom, October 7-10, 1996.

Kujanik S: Teaching Physiology and System of Control of Study Results in Kosice. Local session of the Belgian Physiological Society - Faculty of Medicine, University of Ghent, Belgium, October 20-24, 1996.

Publications Last 3 Years:

1. Mikulecky M Sr., Kujanik S Sr.: Circa- and ultradians in the occurrence of simple extrasystoles in healthy men at lowland in the light of inferential statistics. Acta Physiol. Hung. 2008, 95(3): 321-329. DOI 10.1556/APhysiol.95.2008.3.4
2. Monograph Kujanik S Sr, Snincak M: Zivotosprava a kardiovaskularny system (Life regimen and the cardiovascular system). Publisher Aprilla s.r.o., Kosice 2008, 140 pp., ISBN 978-80-89346-11-0 (In Slovak).
3. Monograph Kujanik S Sr: Niektore aktualne otazky neinvazivnej kardiologie (Some actual topics of non-invasive cardiology). Publisher Aprilla s.r.o., Kosice 2009, 203 pp., ISBN 978-80-89346-14-1 (In Slovak).
4. Kujanik S Sr: Niektore aktualne otazky neinvazivnej kardiologie II (Some actual topics of non-invasive cardiology II). Publisher Aprilla s.r.o., Kosice 2009, 226 pp., ISBN 978-80-89346-20-2 (In Slovak).
5. Kujanik S, Mikulecky M: Circadian and ultradian extrasystole rhythms in healthy individuals at elevated versus lowland altitudes. Int J Biometeorol, 2010, 54(5): 531-538. DOI 10.1007/s00484-010-0302-8.
6. Kujanik Sr S, Mikulecky Sr M: Circa(ultra)dians of single extrasystoles in chronic respiratory failure versus health at lowland. Bratisl Lek Listy, 2010, 111(6): 321-324.

7. Kujanik S Sr, Bacova I.: Spravny pitny rezim (The correct drinking regimen). Slov. Lek., 2008, 18(1-2), pp. 20-26. (In Slovak).
8. Kujanik S Sr, Snincak M: Vysoky vyskyt neskorych komorovych potencialov u chronickej respiracnej insuficiencie vychodoslovenskeho obyvatelstva - prve vysledky (High occurrence of late ventricular potentials in chronic respiratory insufficiency of Eastern Slovakia residents - the first results). Slov. Lek., 2008, 18(3-4), pp. 38-42. (In Slovak).
9. Kujanik S Sr: Potrava a kardiovaskularny system (Food and the cardiovascular system). Slov. Lek., 2008, 18(3-4), pp. 67-74. (In Slovak).
10. Kujanik S Sr: Tuky a kardiovaskularny system (Lipids and the cardiovascular system). Slov. Lek., 2008, 18(7-8), pp. 150-155. (In Slovak).
11. Kujanik S Sr: Hypertenzia: Meriame spravne krvny tlak? (Hypertension: Do we measure the blood pressure accurately?) Slov. Lek., 2008, 18(9-10), pp. 182-189. (In Slovak).
12. Kujanik S Sr, et al.: Pohlavne rozdiely neskorych komorovych potencialov u chronickych respiracnych ochoreni vychodoslovenskeho obyvatelstva - prve vysledky (Gender differences of late ventricular potentials in chronic respiratory diseases of Eastern Slovakia residents - the first results). Slov. Lek., 2008, 18(11-12), pp. 212-216. (In Slovak).
13. Kujanik S Sr: Biologicke rytmy a kardiovaskularny system (Biological rhythms and the cardiovascular system). Slov. Lek., 2009, 19(3-4), pp. 47-53. (In Slovak).
14. Co-author of the textbook MEDICAL PHYSIOLOGY for Slovak students of medicine. Osveta, Martin 2009, 743 pp. (In Slovak).
15. Kujanik S Sr: Inzulin, hyperglykemia a aterogeneza (Insulin, hyperglycaemia and atherogenesis). Slov. Lek., 2009, 19(9-10), pp. 193-199. (In Slovak).
16. Kujanik S Sr: Korelacia elektrickej systoly a srdcovych dysrytmii u chronickeho respiracneho zlyhavania (Correlation of the electric systole with cardiac dysrhythmias in chronic respiratory insufficiency). Slov. Lek., 2009, 19(Supl 1), pp. 2-6. (In Slovak).
17. Kujanik S Sr, Snincak M: Neskore komorove potencialy pri chronickych respiracnych ochoreniach vychodoslovenskeho obyvatelstva - pilotna studia (Late ventricular potentials in chronic respiratory diseases of Eastern Slovakia residents - the pilot study). Slov. Lek., 2009, 19(Supl 1), pp. 7-11. (In Slovak).
18. Kujanik S Sr, Snincak M: Neskore komorove potencialy u vychodoslovenskych pacientov s chronickou bronchitidou - prve vysledky (Late ventricular potentials in Eastern Slovakia patients with chronic bronchitis - the first results). Slov. Lek., 2009, 19(Supl 1), pp. 12-15. (In Slovak).
19. Kujanik S Sr, Snincak M: Vyssi vyskyt neskorych komorovych potencialov u mladych zdravych ludi Vychodneho Slovenska - pilotna studia (Higher occurrence of late ventricular potentials in young healthy persons of Eastern Slovakia - the pilot study). Slov. Lek., 2009, 19(Supl 1), pp. 16-19. (In Slovak).
20. Kujanik S Sr: Cirkaseptanny rytmus v medicine a kardiologii (Circaseptan rhythm in medicine and cardiology). Slov. Lek., 2010, 20(7), pp. 120-127. (In Slovak).

In: Anatomy and Physiology Research
Editors: N. R. Zirena and X. B. Liu
ISBN: 978-1-62942-595-8

Chapter 15

Biographical Sketch - Anatomy and Physiology Research

G. Y. Levin

Affiliation:

Federal State Institution Nizhniy Novgorod Research Institute of Traumatology and Orthopedics

Address:

110 B, Gagarin Avenue, apt.12, Nizhniy Novgorod, Russia, 603009

Date of Birth:

September 6, 1939

Education:

Medical Institute
Department of Health and Social Development of Russia, Nizhniy Novgorod, Russia

Research and Professional Experience:

40 years (haemothology, pathophysiology, burns, microcirculation, haemoreology, plasmapheresis) Professional Appointments: Head of Department of Gravitation Surgery and Hemodialysis

Professional Appointments:

Head of Department of Gravitation Surgery and Hemodialysis

Honors:

Professor, Honoured Scientist

Publications Last 3 Years:

1. Levin G.Ya., Yegorikhina M.N. Role of peroxidation in the aggregation of blood cells in burn disease // Clinical laboratory diagnosis 2008 8. P.43-44.
2. Levin G.Ya., Egorihina M.N. Oxidative stress and aggregation of erythrocytes in burn disease // Revista ozonoterapia 2009. Vol.3, â„1.P.37-40.
3. Levin G.Ya., Egorihina M.N. The role of fibrinogen in aggregation of platelets in burn injury // Burns. 2010 Vol.36, â„ â„6. P. 806-810.
4. Egorikhina M.N., Levin G.Ya. Fibrinogen and Aggregation of Thrombocytes (on the model of thermal trauma) // Yaroslavl pedagogical bulletin.
5. 2010 Vol.3 (Natural sciences), â„ 4 P.61-61-66.

In: Anatomy and Physiology Research
Editors: N. R. Zirena and X. B. Liu

ISBN: 978-1-62942-595-8

Chapter 16

Biographical Sketch - Anatomy and Physiology Research

Marija Meznaric

Affiliation:

MD, Neurologist

Address:

University of Ljubljana, Medical faculty, Institute of Anatomy, Korytkova 2, SI-1000 Ljubljana, Slovenia

Date of Birth:

March 17, 1956

Education:

MD, Neurologist

Research and Professional Experience:

Neuromuscular diseases (muscular dystrophies and mitochondrial myopathies); teaching anatomy and neuroanatomy, reader of muscle biopsies

Professional Appointments:

Assistant at the Institute of Anatomy, Medical faculty, University of Ljubljana; Biobanker (Slovenian partner in the network EuroBioBank

Publications Last 3 Years:

Meznaric, Marija, Gonzalez-Quereda, Lidia, Gallardo, Eduard, Gallano, Pia, De Luna, Noemi, Fanin, Marina, Angelini, Corrado, Peterlin, Borut, Zidar, Janez. Abnormal expression of dysferlin in skeletal muscle and monocytes supports primary dysferlinopathy in patients with one mutated allele. Eur. j. neurol., 2011, vol. 18, issue 7, str. 1021-1023, doi: 10.1111/j.1468-1331.2010.03240.x.

Rogaäœ, Mihael, Meznaric, Marija, Zeviani, Massimo, Sperl, Wolfgang, Neubauer, David. Functional outcome of children with mitochondrial diseases. Pediatr. neurol.. [Print ed.], 2011, vol. 44, no. 5, str. 340-346, doi: 10.1016/j.pediatrneurol.2010.12.004.

Spinazzola, A., Invernizzi, F., Carrara , F., Lamantea, E., Donati, A., Dirocco, M., Giordano, I., Meznaric, Marija, Baruffini, E., Ferrero, I. , Zeviani, M. Clinical and molecular features of mitochondrial DNA depletion syndromes. J. inherit. metab. dis., 2009, letn. 32, št. 2, str. 143-158, doi: 10.1007/s10545-008-1038-z.

Meznaric, Marija, Kralj, Eduard, Angelini, Corrado, Fanin, Marina, Trinkaus, Darinka. Cardiomyopathy in a patient with limb-girdle muscular dystrophy type 2D : pathomorphological aspects. V: Balažic, Jože (ur .). Proceedings of the 17th Intternational Meeting on Forensic Medicine Alpe-Adria-Panonia, Portorož, Slovenia, May 7-10, 2008, (Forensic Science International Supplement Series, vol. 1, iss. 1). [Shannon]: Elsevier, 2009, 2009, letn. 1, št. 1, suppl., str. 58-62.

Rodi, Zoran, Filipoviäœ, Tatjana, Svigelj, Viktor, Meznaric, Marija, Zidar, Janez. Critical illness myopathy in patients with central nervous system disorders = Miopatija kritiÄ• ne bolezni pri bolnikih z okvaro osrednjega živÄ□evja. Zdrav Vestn, 2009, letn. 78, št. 6/7, str. 289-293.

In: Anatomy and Physiology Research
Editors: N. R. Zirena and X. B. Liu
ISBN: 978-1-62942-595-8

Chapter 17

Biographical Sketch - Anatomy and Physiology Research

Seung-June Oh

Address:

Tel: +82-2-2072-2406 Fax: +82-2-742-4665. e-mail: sjo@snu.ac.kr

Date of Birth:

1965

Education:

Mar. 1997 - Feb. 1999 - Ph. D. degree; Urology Major, College of Medicine, Seoul National University, Seoul, Korea; Thesis: The mechanism and characteristics of urinary bladder smooth muscle contractions: comparative study in neonatal and adult rats

Mar. 1991 - Feb. 1993; Master degree; Urology Major, College of Medicine, Seoul National University, Seoul, Korea;
Thesis: The effect of intravesical oxybutynin instillation, on rat voiding cycle

Mar. 1983 - Feb. 1989 - Bachelor degree; Seoul National University College of Medicine

Research and Professional Experience:

Benign prostatic hyperplasia, Neurogenic bladder, Neurourology,
Urinary incontinence, Overactive bladder,
Quality of Life Research, Reconstructive Surgery

Professional Appointments:

Sep. 1, 2010 - present, Professor, Department of Urology, Seoul National University College of Medicine
Oct. 2005 - Aug. 2010, Associate Professor, Department of Urology, Seoul National University College of Medicine
Sep. 2000 - Sep. 2005, Assistant Professor, Department of Urology, Seoul National University College of Medicine
Sep. 1, 2010 - present, Professor and Physician, Department of Urology, Seoul National University Hospital
Sep. 2000 - Aug. 2010, Professor, Department of Urology, Seoul National University Hospital

Honors:

1. May 15, 2010. Academic Award, Best International Paper of the Year Category, Korean Continence Society
2. Nov 20, 2009. Best Paper Award, Korean Urological Association
3. Nov 20, 2009. Best Video Award, Korean Urological Association
4. Apr 19, 2006. Young Researcher Award, Seoul National University Hospital/ Seoul National University College of Medicine
5. Oct 28, 2006. Jeil Academic Award (Best Paper of the Year Award), Korean Continence Society

Publications Last 3 Years:

INTERNATIONAL BOOK CHAPTERS

1. Ku JH, Oh SJ. Comparison of Three Quality of Life Questionnaires in Urinary Incontinence. In: Preedy, Victor R.; Watson, Ronald R. Editors. Handbook of Disease Burdens and Quality of Life Measures, Springer (4448 pages. in 6 volumes) eReference (online access, http://www.springerlink.com/content/hw6vn9875m0nv985/); 2010
2. Oh SJ, Stoffel JT, McGuire EJ. Pubovaginal sling. In: Wein AJ, Kavoussi LR, Novick AC, Partin AW, Peters CA, editors. Campbell-Walsh urology, 9th ed. Philadelphia: Saunders; 2007;2234-50.

DOMESTIC BOOK CHAPTERS (written in Korean)

1. Oh SJ. Classification of neurogenic bladder. In: Korean Continence Society, editor. Textbook of Voiding Dysfunction and Female Urology. 2nd edition, Ilchokak, Seoul, 2009;107-112. (in Korean)
2. Oh SJ, Shin HS. Voiding Dysfunction, In: Han TR, Bang MS, editors. Principle of Rehabilitation Medicine, version 3. Koonja, Seoul 2008:429-51. (in Korean)

3. Oh SJ, Lim SH. Conservative Treatment Program, In: Korean Continence Society, editor. Conservative Management of Female Incontinence. Pentaid, Seoul, 2007;177-88. (in Korean)

INTERNATIONAL JOURNAL
(Science Citation Index or SCI Expanded)

1. Li WJ, Shin MK, Oh SJ. Time Dependent Bladder Apoptosis Induced by Acute Bladder Outlet Obstruction and Subsequent Emptying is Associated with Decreased MnSOD Expression and Bcl-2/Bax Ratio. J Korean Med Sci. 2010 Nov;25(11):1652-6. Epub 2010 Oct 26. [Principal Investigator]
2. Jeong SJ, Cho SY, Oh SJ. Spinal cord/brain injury and the neurogenic bladder. Urol Clin North Am. 2010 Nov;37(4):537-46. [Principal Investigator]
3. Li WJ, Oh SJ. Diabetic cystopathy is associated with PARP/JNK/mitochondrial apoptotic pathwaymediated bladder apoptosis. Neurourol Urodyn. 2010 Sep;29(7):1332-7. [Principal Investigator]
4. Park YH, Ku JH, Oh SJ. Accuracy of post-void residual urine volume measurement using a portable ultrasound bladder scanner with real-time pre-scan imaging. Neurourol Urodyn. 2010 Jul 23. [Epub ahead of print] [Principal Investigator]
5. Kang MY, Ku JH, Oh SJ. Non-invasive parameters predicting bladder outlet obstruction in Korean men with lower urinary tract symptoms. J Korean Med Sci. 2010 Feb;25(2):272-5. [Principal Investigator]
6. Ku JH, Ko DW, Cho JY, Oh SJ. Correlation Between Prostatic Urethral Angle and Bladder Outlet Obstruction Index in Patients With Lower Urinary Tract Symptoms. Urology. 2010 Jun;75(6):1467-71. Epub 2009 Dec 4. [Principal Investigator]
7. Ku JH, Cho SY, Oh SJ. Residual fraction as a parameter to predict bladder outlet obstruction in men with lower urinary tract symptoms. Int J Urol. 2009 Sep;16(9):739-44. Epub 2009 Jul 30. [Principal Investigator]
8. Cho SY, Oh SJ. The clinical significance of rectal contractions that occur during urodynamic studies. Neurourol Urodyn. 2009 Jul 27. [Epub ahead of print] [Principal Investigator]
9. Kong HJ, Park S, Lee T, Lee JY, Kim HC, Oh SJ. Novel natural fill telemetric pressure flow study of discomfort and bladder outlet obstruction. J Urol. 2009 Aug;182(2):601-6. Epub 2009 Jun 17. [Principal Investigator]
11. Kim SH, Kim TB, Kim SW, Oh SJ. Urodynamic findings of the painful bladder syndrome/interstitial cystitis: a comparison with idiopathic overactive bladder. J Urol. 2009 Jun;181(6):2550-4. Epub 2009 Apr [Principal Investigator] 11. Cho SY, Yi JS, Oh SJ. The clinical significance of poor bladder compliance. Neurourol Urodyn. 2009;28(8):1010-4. [Principal Investigator]
12. Oh SJ, Ku JH, Choo MS, Yun JM, Kim DY, Park WH. Health-related quality of life and sexual function in women with stress urinary incontinence and overactive bladder. Int J Urol. 2008 Jan;15(1):62-7; discussion [Principal Investigator]
13. Oh SJ, Choo MS, Kim HS, Kim JC, Lee JG, Yun JM, Kim DY, Paick JS, Lee JY, Chung BS, Min KS, Kim YH, Jung HC, Son H, Jeong JY, Rho J, Lee KS, Park WH, Ku JH. Generic and disease-specific healthrelated quality of life in women with coital

incontinence: a prospective, multicenter study. Gynecol Obstet Invest. 2008;65(1):62-7. Epub 2007 Sep 11. [Principal Investigator]

14. Choo MS, Ku JH, Oh SJ, Lee KS, Paick JS, Seo JT, Kim DY, Lee JJ, Lee JG, Na YG, Kwon DD, Park WH. Prevalence of urinary incontinence in Korean women: an epidemiologic survey. Int Urogynecol J Pelvic Floor Dysfunct. 2007 Nov;18(11):1309-15. Epub 2007 Feb 15.
15. Paick JS, Cho MC, Oh SJ, Kim SW, Ku JH. Is proximal urethral mobility important for transobturator tape procedure in management of female patients with stress urinary incontinence? Urology. 2007 Aug;70(2):246-50; discussion 250-1.
16. Paick JS, Kim SW, Oh SJ, Ku JH. Voiding patterns in men and women with lower urinary tract symptoms combined with nocturia. Int J Urol. 2007 Aug;14(8):699-703.
17. Paick JS, Cho MC, Oh SJ, Kim SW, Ku JH. Factors influencing the outcome of mid urethral sling procedures for female urinary incontinence. J Urol. 2007 Sep;178(3 Pt 1):985-9; discussion 989. Epub 2007 Jul 16.
18. Oh SJ, Son H, Jeong JY, Ku JH. Anticholinergics in patients with overactive bladder: Assessment of ambulatory urodynamics and patient perception. Neurourol Urodyn. 2007;26(6):789-93. [Principal Investigator]
19. Paick JS, Oh SJ, Kim SW, Ku JH. Tension-free vaginal tape, suprapubic arc sling, and transobturator tape in the treatment of mixed urinary incontinence in women. Int Urogynecol J Pelvic Floor Dysfunct. 2008 Jan;19(1):123-9. Epub 2007 Jun 13.
20. 20. Oh SJ, Ku JH. Comparison of three disease-specific quality-of-life questionnaires (Bristol Female Lower Urinary Tract Symptoms, Incontinence Quality of Life and King's Health Questionnaire) in women with stress urinary incontinence. Scand J Urol Nephrol. 2007;41(1):66-71. [Principal Investigator]
21. Paick JS, Cho MC, Oh SJ, Kim SW, Ku JH. Influence of self-perceived incontinence severity on quality of life and sexual function in women with urinary incontinence. Neurourol Urodyn. 2007;26(6):828-35.
22. Oh SJ, Ku JH.Impact of self-perceived bothersomeness on health-related quality of life in women with overactive bladder symptoms. Urol Int. 2007;78(1):63-9. [Principal Investigator]
23. Paick JS, Kim SW, Oh SJ, Ku JH. A generic health-related quality of life instrument, the Medical Outcomes Study Short Form-36, in women with urinary incontinence. Eur J Obstet Gynecol Reprod Biol. 2007 Jan;130(1):18-24.
24. Oh SJ, Ku JH. Impact of stress urinary incontinence and overactive bladder on micturition patterns and health-related quality of life. Int Urogynecol J Pelvic Floor Dysfunct. 2007 Jan;18(1):65-71. [Principal Investigator]

In: Anatomy and Physiology Research
Editors: N. R. Zirena and X. B. Liu
ISBN: 978-1-62942-595-8

Chapter 18

Biographical Sketch - Anatomy and Physiology Research

Pedro Fontes Oliveira

Affiliation:

CICSUBI

Address:

CICS UBI (Health Sciences Research Center)
Faculty of Health Sciences
University of Beira Interior
Av. Infante D. Henrique
6200506 Covilhr
Portugal

Date of Birth:

29/08/1973

Education:

2008 - Postdoctoral in Biological Sciences
2004 – PhD; Biomedical Sciences
1996 - LSc in Biochemistry

Research and Professional Experience:

2009 - Assistant Researcher
2004/2008 – Postdoctoral researcher
1998/2004 – PhD student
1996/1998 – Young scientist researcher

Professional Appointments:

2009/2010 – Invited Professor at FCSUBI, Portugal
2002/2003 – Assistant Professor ate ISAVE, Portugal

Publications Last 3 Years:

Alves, M., Oliveira, P.F., Martins, F.O., Oliveira, P.J., Carvalho, R.A. (2011) Metabolic Adjustments During Ischemia Role of Gender and Cardioplegia. European Journal of Pharmacology (under review)

Rato, L., Alves, M.G., Socorro, S., Carvalho, R.A., Cavaco, J.E., Oliveira, P.F. (2011) Androgens and Estrogens as metabolic modulators in cultured rat Sertoli cells. Reproduction (under review)

Oliveira, P.F., Alves, M.G., Rato, L., Só, R., Barros, A., Sousa, M., Carvalho, R.A., Cavaco, J.E., Socorro, S. (2011) Metabolic influence of 5αdihydrotestosterone and 17βestradiol on human Sertoli cells. Human Reproduction (under review).

Laurentino, S., Correia, S., Cavaco, J.E., Oliveira, P.F., Rato, L., Sousa, M., Barros, A., Socorro, S. (2011) Regucalcin/SMP30 is broadly expressed in reproductive tissues and a new androgen target gene in mammalian testis. Journal of Cellular Biochemistry (under review).

Rato, L., Socorro, S., Cavaco, J., Oliveira, P.F. (2010) Tubular Fluid Secretion in the Seminiferous Epithelium: Ion Transporters and Aquaporins in Sertoli Cells. Journal of Membrane Biology 236(2):21524.

Oliveira, P.F., Sousa, M., Barros, A., Moura, T., Rebelo da Costa, A. (2009) Membrane transporters and cytoplasmatic pH regulation on bovine Sertoli cells. Journal of Membrane Biology 227: 4955.

Oliveira, P.F., Sousa, M., Barros, A., Moura,T., Rebelo da Costa, A. (2009) Intracellular pH regulation in human Sertoli cells: role of membrane transporters. Reproduction 137: 353359.

Oliveira, P.F., Rebelo da Costa, A. and Ferreira H.G. (2008) A mathematical model of ion transport in the outer mantle epithelium of Anodonta cygnea. Journal of Membrane Biology 223 (2): 5972.

In: Anatomy and Physiology Research
Editors: N. R. Zirena and X. B. Liu

ISBN: 978-1-62942-595-8

Chapter 19

Biographical Sketch - Anatomy and Physiology Research

James Kirkby-Bott

Affiliation:

Mr.

Address:

E-mail: james.kirkby1@mac.com; james.kirkby-bott@uhbristol.nhs.uk

Date of Birth:

07/Oct/1973

Education:

Research and Professional Experience:

Research registrar at Imperial College London, Endocrine Fellow Hammersmith Hospital, Imperial College, London, Endocrine Fellowship CHRU, Lille, France

Professional Appointments:

Locum Consultant Endocrine Surgeon, BRI (UHBT) Bristol, UK

Honors:

Norman Tanner Prize Medal in Surgery, RSM, London 2008

Publications Last 3 Years:

J Kirkby-Bott, A Skandarajah, H Markogiannakis, M Cowan, B Fleming, F Palazzo. *Pre-operative vitamin D deficiency predicts post-operative hypocalcaemia after total thyroidectomy*. World J Surg, Accepted for publication 27th October 2010

Johns R, Kirkby-Bott J. *Snapshot of Stewart-Treves syndrome.* Br J Surg, Accepted for publication 14th October 2010.

Kirkby-Bott J, Jones E, Perring S, Hosking S.W. *Proximal acid reflux treated by fundoplication predicts a good outcome for chronic cough attributable to gastro oesophageal reflux disease.* Langenbeck Arch Surg, Accepted for publication 14th July 2010. DOI: 10.1007/s00423-010-0702-6

J.Kirkby-Bott, Z El-Khatib, B Soudan, R Caiazzo, L Arnalsteen, B Carnaille. *25 hydroxy vitamin D deficiency causes parathyroid incidentalomas.* Langenbeck Arch Surg,(2010). 395; 919-924. DOI: 10.1007/s00423-010-0692-4

J.Kirkby-Bott, B.Carnaille. *Experience of the reliability of intra-operative sampling of tissue PTH in parathyroid surgery (letter to the editor).* World J Surg (2010) 34:2254. doi:10.1007/s00268-010-0615-0

J.Kirkby-Bott, M Nelson. *Consultant Recruitment is Falling while Trainee numbers are increasing: A Worrying Trend over a seven year period.* Bulletin of the Royal College of Surgeons, England 2010; 92(6): 1-4

R Caiazzo, L Arnalsteen, M Pigeyre, J Kirkby-Bott, A Severine, P Pigny, P Mathourin, M Romon, P Fontaine, F Pattou. *Metabolic outcomes and quality of life 5 years after Laparoscopic Adjustable Gastric Band in Grade 3 obese patients with type 2 diabetes mellitus.* Br J Surg 2010; 97: 884-891

J.Kirkby-Bott, A.Brent. *'Open Surgical insertion of Gastric feeding tube under local anaesthesia – a technical strategy'* Ann R Coll Surg 2008; 90: 696–703

In: Anatomy and Physiology Research
Editors: N. R. Zirena and X. B. Liu
ISBN: 978-1-62942-595-8

Chapter 20

Biographical Sketch - Anatomy and Physiology Research

Tim Palmer

Affiliation:

BSc(Hons) PhD

Date of Birth:

15/6/1966

Education:

1984 - 1987 - BSc(Hons); Biochemistry,
University of Manchester, UK
1987 - 1991 – PhD; Biochemistry,
University of Glasgow, UK

Research and Professional Experience:

1991-1996: Postdoctoral R.A. and 1996-1997 Aha Postdoctoral Fellow, Cardiology Division, Duke University Medical Centre, North Carolina, USA

1997-2004, Lecturer, 2004-2008 Senior Lecturer, 2008-Present Reader, Institute of Cardiovascular and Medical Sciences, University of Glasgow, UK

Publications Last 3 Years:

Rutherford C, Palmer TM. Molecular basis of protective anti-inflammatory signalling by cyclic AMP in the vascular endothelium, in "Systems Biology of Signaling Networks" (Editor Sangdun Choi). 561-588, Springer, 2010, ISBN: 978-1-4419-5796-2.

Safhi MMA, Rutherford C, Ledent C, Sands WA, Palmer TM. Priming of signal transducer and activator of transcription proteins for polyubiquitylation and degradation by the A_{2A} adenosine receptor. Mol. Pharmacol. 2010 77:968–978.

Mundell SJ, Matharu A-L, Palmer TM, Benovic JL, Kelly E. Deletion of the distal COOH-terminus of the A_{2B} adenosine receptor switches agonist-induced internalisation to an arrestin- and clathrin-independent pathway and inhibits receptor recycling. Br J Pharmacol 2010 159:518–533.

Woolson HD, Thompson VS, Rutherford C, Yarwood SJ, Palmer TM. Selective inhibition of cytokine-activated extracellular signal-regulated kinase by cyclic AMP via Epac1-dependent induction of suppressor of cytokine signalling-3. Cell. Signal. 2009 21:1706–1715.

Borland G, Bird R, Palmer TM, Yarwood SJ. Activation of protein kinase C α by EPAC1 is required for the ERK- and C/EBPβ-dependent induction of the SOCS-3 gene by cyclic AMP in COS1 cells. J. Biol. Chem. 2009 284:17391–17403.

Yarwood SJ, Borland G, Sands WA, Palmer TM. Identification of C/EBPs as EPAC-activated transcription factors that mediate the induction of the SOCS-3 gene. J Biol Chem. 2008 283(11):6843-53.

Mair KM, MacLean MR, Morecroft I, Dempsie Y, Palmer TM. Novel interactions between the 5-HT transporter, 5-HT_{1B} receptors and Rho kinase in vivo and in pulmonary fibroblasts. Br J Pharmacol. 2008 155:606-616.

Palmer TM, Trevethick MA. Suppression of inflammatory and immune responses by the A(2A) adenosine receptor: an introduction. Br J Pharmacol. 2008 153:S27–S34.

Sands WA, Palmer TM. Regulating gene transcription in response to cyclic AMP elevation. Cell Signal. 2008 Mar;20(3):460-6.

In: Anatomy and Physiology Research
Editors: N. R. Zirena and X. B. Liu

ISBN: 978-1-62942-595-8

Chapter 21

Biographical Sketch - Anatomy and Physiology Research

José Ríos-Díaz

Affiliation:

Dpt. Health Sciences (Physical Therapy). UCAM. Murcia (Spain)

Address:

UCAM. Campus Los Jerónimos s/n. 30107- Guadalupe-Murcia-Spain

Date of Birth:

1973 Nov 28th

Education:

Doctor in Biomedical Sciences, Biologist and Physiotherapist

Research and Professional Experience:

Director of "Preventive Ultrasonography and Morpho-densitometry" Research Group.

Titular teacher of Human Physiology, Physical Therapy Fundamentals and Research Metodology.

Publications Last 3 Years:

1. Martínez Payá JJ, Palomino Cortés MA, Ríos-Díaz J, Del Baño Aledo ME. Influencia morfológica del tendón de Aquiles sobre la densidad mineral ósea del calcáneo. In: Roca de Togores Muñoz C, Rodes Lloret F, eds. Antropología Física y Forense. Alicante: Instituto Alicantino de Cultura Juan Gil-Albert, Diputación de Alicante (Spain); 2007. p.195-202.
2. Martínez-Almagro A, Aleixandre Benavent R, Fernández Aparicio T, Ríos-Díaz J, Navarro Coy M. Terminología, método científico y estadística aplicada en ciencias de la salud. Murcia (Spain): Morphos Ediciones S.L.; 2007. ISBN: 978-84-611-57.
3. Ríos-Díaz J, Martínez Payá JJ, Palomino Cortés MA, Del Baño Aledo ME. Dimensión fractal del hueso esponjoso de calcáneo. In: Roca de Togores Muñoz C, Rodes Lloret F, eds. Antropología Física y Forense. Alicante: Instituto Alicantino de Cultura Juan Gil-Albert, Diputación de Alicante (Spain); 2007. p.187-1 93.
4. Martínez Payá JJ, Ríos-Díaz J, Martínez Pérez LM, Montaño Munuera JA, Peña Marín O. Bases físicas y morfológicas para la interpretación de la imagen ecográfica músculo-esquelética. In: Martínez Payá JJ, editor. Anatomía ecográfica del hombro. Madrid (Spain): Editorial Médica Panamericana; 2008. p.3-38. ISBN: 978-84-9835-054-8.
5. Ríos-Díaz J, Martínez-Payá JJ, Palomino Cortés MA, del Baño Aledo ME, Pérez Hernández M. Análisis morfológico-estructural del hueso trabecular de la porción proximal del fémur mediante las variables de dimensión fractal, lagunaridad y textura. In: Nieto Amada JL, Obón Nogués JA, Baena Pinilla S, editores. Genes, ambiente y enfermedades en poblaciones humanas. Zaragoza (Spain): Prensas Universitarias de Zaragoza; 2008. p. 341-356. ISBN: 978-84-92521 -49-4.
6. Meroño Gallut AJ, Montaño Munuera JA, Zamora Conesa V, Martínez Fuentes J, Murcia González AM, Ríos-Díaz J, Buendía López F. VII Jornadas Interuniversitarias de Fisioterapia. Murcia (Spain): UCAM editorial; 2009. ISBN: 84-96353796.
7. Ríos-Díaz J, Martínez-Fuentes J, Meroño-Gallut AJ, Pérez-Llanes R, Castillo Lozano R, Alcantarilla Pedrosa M. Proyecto de innovación educativa para el desarrollo de competencias en la materia Trabajo Fin de Grado de los estudios de Fisioterapia. Congreso Internacional de Innovación Docente. Cartagena. Murcia (Spain): 2011. ISBN: 978-84-694-5333-9.
8. De la Cruz Torres B, Albornoz Cabello M, Cruz Sicilia S, Meroño-Gallut AJ, RíosDíaz J, Martínez-Fuentes J. La metodología ABP como innovación docente en la Facultad de Fisioterapia. Congreso Internacional de Innovación Docente. Cartagena. Murcia (Spain): 2011. ISBN: 978-84-694-5333-9.
9. Montoya Carralero MD, Ríos-Díaz J, Martínez Fuentes J. Prevención de las úlceras por presión en pacientes terminales e inmovilizados seguidos por el equipo de soporte de atención domiciliaria. Rev Esp Geriatr Gerontol 2007;42(5):263-70. (ISSN: 021 1-139X).
10. Del Baño Aledo ME, Martínez Payá JJ, Ríos-Díaz J, Palomino Cortés MA. Aplicación en fisioterapia de la valoración cuantitativa de las características morfoecogénicas del tendón de Aquiles. Fisioterapia. 2008;30(2):61-68. (ISSN: 0211-5638).

11. **Ríos-Díaz J**, Linares Hevilla FJ, Martínez-Payá JJ, Palomino Cortés MA, del Baño Aledo ME. Arquitectura y organización interna del hueso ante la aplicación de diferentes estímulos mecánicos. Fisioterapia. 2008;30(4): 194-203. (ISSN: 0211-5638).
12. **Ríos-Díaz J**, Martínez-Payá JJ, del Baño Aledo ME. El análisis textural mediante las matrices de co-ocurrencia (GLCM) sobre imagen ecográfica del tendón rotuliano se muestra útil para detectar cambios histológicos tras un entrenamiento con plataforma de vibración. Cult cienc deporte. 2009;4(11):91-102. (ISSN: 1696-5043).
13. **Ríos-Díaz J**, Caeiro Rey JR, Martínez Payá JJ, Dapía Robleda S, del Baño Aledo ME. Relación entre las variables morfo-estructurales de dimensión fractal, lacunaridad y entropía en el hueso trabecular de la porción proximal del fémur. Rev Esp Enferm Metab Óseas. 2009;18(1):2-8. (ISSN: 1132-8460).
14. **Ríos-Díaz J**, Martínez Payá JJ, de Groot Ferrando A, del Baño Aledo ME. Nuevo método de análisis textural mediante matrices de co-ocurrencia del nivel de gris sobre imagen ecográfica del tendón de Aquiles: diferencias entre deportistas y sedentarios. Cuest Fisioter. 2009;38(2):68-79. (ISSN: 1135-8599).
15. **Ríos-Díaz J**, De-Groot-Ferrando A, Martínez-Payá JJ, Del-Baño-Aledo ME. Fiabilidad y reproducibilidad de un nuevo método de análisis morfo-textural de imágenes ecográficas del tendón rotuliano. Reumatol Clin. 2010;6(6):278-84. (ISSN: 1699-258X). Índice impacto nacional (IME-CSIC) de 0,235.
16. Martínez-Fuentes J, Meroño Gallut AJ, **Ríos-Díaz J**. El factor de impacto como criterio para la evaluación de la producción y la calidad científica. Rev Iberoam Fisioter Kinesiol.201 0;1 3(1):29–36. (ISSN: 1138-6045).
17. **Ríos-Díaz J**, Martínez-Payá JJ, Del-Baño-Aledo ME, De-Groot-Ferrando A. Fiabilidad y reproducibilidad intra e inter observador de un método semi-automático de análisis ecográfico del tendón de Aquiles. Cuest fisioter. 2010; 39(3): 190-198. (ISSN: 0033-8338).
18. **Ríos-Díaz J**, Martínez-Payá JJ, del-Baño-Aledo ME, de-Groot-Ferrando A, PérezLlanes R. Análisis discriminante del patrón textural ecográfico con matrices de concurrencia como nueva herramienta para el estudio del tendón. Fisioterapia. 2011. In Press. (ISSN: 0211-5638).
19. Pérez-Llanes R, **Ríos-Díaz J**, Martínez-Payá JJ, del-Baño-Aledo ME. Análisis ecográfico de las modificaciones en la velocidad del flujo sanguíneo en la arteria vertebral producidas por el movimiento de rotación cervical máxima. Fisioterapia. 2011. In Press.

In: Anatomy and Physiology Research
Editors: N. R. Zirena and X. B. Liu

ISBN: 978-1-62942-595-8

Chapter 22

Biographical Sketch - Anatomy and Physiology Research

Norihisa Saeki

Affiliation:

M.D., Ph.D

Address:

5-1-1 Tsukiji, Chuo-ku, Tokyo 104-0045, Japan

Date of Birth:

June 16, 1966

Education:

School of Medicine,
University of the Ryukyus, Okinawa, Japan
Graduate School of Medical Science,
University of the Ryukyus, Okinawa, Japan

Professional Appointments:

Senior Staff Scientist

Publications Last 3 Years:

1. Saeki N, Saito A, Choi IJ, Matsuo K, Ohnami S, Totsuka H, Chiku S, Kuchiba A, Lee YS, Yoon KA, Kook MC, Park SR, Kim YW, Tanaka H, Tajima K, Hirose H, Tanioka F, Matsuno Y, Sugimura H, Kato S, Nakamura T, Nishina T, Yasui W, Aoyagi K, Sasaki H, Yanagihara K, Katai H, Shimoda T, Yoshida T, Nakamura Y, Hirohashi S, Sakamoto H. A functional SNP in *MUC1*, at chromosome 1q22, determines susceptibility to diffuse-type gastric cancer. *Gastroenterology* (2011)
2. Wang K, Diskin SJ, Zhang H, Attiyeh EF, Winter C, Hou C, Schnepp RW, Diamond M, Bosse K, Mayes PA, Glessner J, Kim C, Frackelton E, Garris M, Wang Q, Glaberson W, Chiavacci R, Le Nguyen, Jagannathan J, Saeki N, Sasaki H, Grant SF, Iolascon A, Mosse YP, Cole KA, Li H, Devoto M, McGrady PW, London WB, Capasso M, Rahman N, Hakonarson H, Maris JM. Integrative genomics identifies LMO1 as a neuroblastoma oncogene. *Nature* 469, 216–220 (2011)
3. Saeki N, Gu J, Yoshida T, Wu X. Prostate Stem Cell Antigen: A Jekyll and Hyde Molecule? *Clin Cancer Res* 16, 3533-3538 (2010) (review)
4. Yoshida T, Ono H, Kuchiba A, Saeki N, Sakamoto H. Genome-wide germline analyses on cancer susceptibility and GeMDBJ database: Gastric cancer as an example. *Cancer Sci* 101, 1582-1589 (2010) (review)
5. Wu X, Ye Y, Kiemeney LA, Sulem P, Rafnar T, Matullo G, Seminara D, Yoshida T, Saeki N, Andrew AS, Dinney CP, Czerniak B, Zhang ZF, Kiltie AE, Bishop DT, Vineis P, Porru S, Buntinx F, Kellen E, Zeegers MP, Kumar R, Rudnai P, Gurzau E, Koppova K, Mayordomo JI, Sanchez M, Saez B, Lindblom A, de Verdier P, Steineck G, Mills GB, Schned A, Guarrera S, Polidoro S, Chang SC, Lin J, Chang DW, Hale KS, Majewski T, Grossman HB, Thorlacius S, Thorsteinsdottir U, Aben KK, Witjes JA, Stefansson K, Amos CI, Karagas MR, Gu J. Genetic variation in the prostate stem cell antigen gene *PSCA* confers susceptibility to urinary bladder cancer. *Nat Genet* 41, 991-995 (2009)
6. Saeki, N., Usui T, Aoyagi K, Kim DH, Sato M, Mabuchi T, Yanagihara K, Ogawa K, Sakamoto H, Yoshida T, Sasaki H. Distinctive expression and function of four *GSDM* family genes (*GSDMA-D*) in normal and malignant upper gastrointestinal epithelium. *Genes Chromosomes Cancers* 48, 261-271(2009)
7. Ohta, H., Aoyagi, K., Fukaya, M., Danjoh, I., Isohata, N, Saeki N, Taniguchi H, Sakamoto H, Shimoda T, Tani T, Yoshida T, Sasaki H. Cross talk between hedgehog and epithelial-mesenchymal transition pathways in gastric pit cells and in diffuse-type gastric cancers. *Br J Cancer* 100, 389-398 (2009)
8. Sakamoto H*, Yoshimura K*, Saeki N*, Katai H, Shimoda T, Matsuno Y, Saito D, Sugimura H, Tanioka F, Kato S, Matsukura N, Matsuda N, Nakamura T, Hyodo I, Nishina T, Yasui W, Hirose H, Hayashi M, Toshiro E, Ohnami S, Sekine A, Sato Y, Totsuka H, Ando M, Takemura R, Takahashi Y, Ohdaira M, Aoki K, Honmyo I, Chiku S, Aoyagi K, Sasaki H, Ohnami S, Yanagihara K, Yoon KA, Kook MC, Lee YS, Park SR, Kim CG, Choi IJ, Yoshida T, Nakamura Y, Hirohashi S. Genetic variation in *PSCA* is associated with susceptibility to diffuse-type gastric cancer. *Nat Genet* 40, 730-740 (2008).

In: Anatomy and Physiology Research
Editors: N. R. Zirena and X. B. Liu

ISBN: 978-1-62942-595-8

Chapter 23

Biographical Sketch - Anatomy and Physiology Research

Kunihiro Sakuma

Affiliation:

Research Center for Physical Fitness, Sports and Health, Toyohashi University of Technology

Date of Birth:

February 18, 1969

Education:

April, 1984 graduated March, 1987 - Entered Fukushima Prefectural Haramachi Senior High School
April, 1987 graduated March, 1991 - Bachelor of Physical Education. Obtained an M.S.; Entered Faculty of Physical Education, University of Tsukuba
March, 1993 - Master of Physical Education
March, 1996 - Ph.D. Doctor of Physical Science; University of Tsukuba for a thesis entitled Several factors inducing region-specific changes in fiber types along the muscle length of ratsâ

Professional Appointments:

Associate Professor

Honors:

Selection of 2000 Outstanding Intellecturals of the 21st Century Awards Programme
2010-2011 - Selection of Marquis Who's Who in the Worl
2009-2010 - Selection of Marquis Who's Who in the World
2011 - Research Grant-in-Aid for Scientific Research C (No. 23500578)
2008-2010 - Research Grant-in-Aid for Scientific Research C (No. 20500575)
2005-2006 - Research Grant-in-Aid for Young Scientists B(No. 17700500)
2003-2004 - Research Grant-in-Aid for Young Scientists B(No. 15700423)
2001-2002 - Research Grant-in-Aid for Encouraging Research (No. 13780029)
1999-2000 - Research Grant-in-Aid for Encouraging Research (No. 11780059) from the Ministry of Education, Culture, Sports, Science and Technology of Japan

Publications Last 3 Years:

1. Sakuma K, Yamaguchi A: Cyclosporin A inhibits the growth of neonatal MHC-expressing myotubes independent of NFATc1 and NFATc3 in the mechanically overloaded soleus muscle of mice. Am J Mol Biol 1: 7-16, 2011.
2. Hokari F, Kawasaki E, Sakai A, Koshinaka K, Sakuma K, Kawanaka K: Muscle contractile activity regulates Sirt3 protein expression in rat skeletal muscles. J Appl Physiol 109: 332-340, 2010.
3. Akiho M, Nakashima H, Sakata M, Yamasa Y, Yamaguchi A, Sakuma K : Expression profile of Notch-1 in mechanically overloaded plantaris muscle of mice. Life Sci 86: 59-65, 2010.
4. Sakuma K, Watanabe K, Hotta N, Koike T, Ishida K, Katayama K, Akima H :Â The adaptive responses in several mediators linked with hypertrophy and atrophy of skeletal muscle after lower limb unloading in humans. Acta Physiol (Oxf) 197:151-159, 2009.
5. Tokugawa S, Sakuma K, Fujiwara H, Hirata M, Oda R, Morisaki S, Yasuhara M,Â Kubo T: The expression pattern of PKCtheta in satellite cells of normal and regenerating muscle in the rat. Neuropathology 29: 211-218, 2009. Review.
6. Sakuma K, Yamaguchi A: Molecular determinants of skeletal muscle hypertrophy. Journal of Sports Medicine & Doping Studies, in press, 2011.
7. Sakuma K, Yamaguchi A: Current application of cyclosporine A to investigate skeletal muscle adaptation. Curr Enzyme Inhibit, in press, 2011.
8. Sakuma K, Yamaguchi A: The recent understanding of the neurotrophin's role in skeletal muscle adaptation. J Biomed Biotechnol 2011, Article ID 201696, 12 pages.
9. Sakuma K, Yamaguchi A: Inhibitors of myostatin- and proteasome-dependent signaling for attenuating muscle wasting. Recent Pat Regen Med 1: 284-298, 2011.
10. Sakuma K, Yamaguchi A: The functional role of calcineurin in hypertrophy, regeneration, and disorders of skeletal muscle. J Biomed Biotechnol 2010: Article ID 721219, 8 pages, 2011.
11. Sakuma K, Yamaguchi A: Molecular mechanisms in aging and currentÂ strategies to counteract sarcopenia. Curr Aging Sci 3: 90-101, 2010.

12. Book Chapter 1. Sakuma K, Yamaguchi A: Sarcopenia: Molecular mechanisms and current therapeutic strategy. Eds Perloft JW and Wong AH, In: Cell Aging (in press), Nova Science Publishers, NY,Â 2011.
13. SakumaK, Yamaguchi A: Serum response factor(SRF)-dependent pathway: Potential mediators of growth, regeneration, and hypertrophy of skeletal muscle. Eds. Pandalai S, In; Recent Research Developments in Life Science (in press), Research Signpost, Kerala, India, 2011.
14. Yamaguchi A, Sakuma K, Morita I: Local and circulating insulin-like growth factor-I on skeletal muscle. Eds Sakuma K, In: Current Research Trends in Skeletal Muscle (in press), Research Signpost, Kerala, India, 2011.
15. Sakuma K, Yamaguchi A: Calcineurin-dependent signaling: potent regulators for fiber-type transformation, regeneration, and hypertrophy of skeletal muscle. Eds Sakuma K, In: Current Research Trends in Skeletal Muscle (in press), Research Signpost, Kerala, India, 2011.

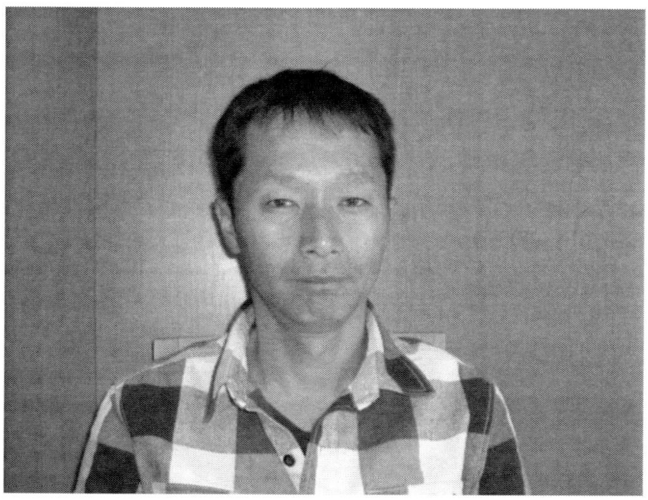

In: Anatomy and Physiology Research
Editors: N. R. Zirena and X. B. Liu

ISBN: 978-1-62942-595-8

Chapter 24

Biographical Sketch - Anatomy and Physiology Research

Anna Naidenova Tolekova

Affiliation:

Associate Professor

Address:

Office address: Department of physiology, pathophysiology and pharmacology, section of physiology, Trakia University, Medical Faculty, Armeiska str. 11, Stara Zagora, 6000, Bulgaria
Home address: 17 Mladost str, B, ap.26, Stara Zagora, 6014, Bulgaria

Date of Birth:

18/08/1955

Education:

1979 - Medical doctor; Higher medical institute "Ivan P. Pavlov", Plovdiv
1988 - Medical Faculty, Sofia, speciality of physiology;
2004 – PhD; VAK

Research and Professional Experience:

Major Fields of scientific research

- Renin-angiotensin system and its regulation
- Modeling of physiological processes
- Role of reactive oxygen species in normal and pathological conditions
- Peptidergic regulation of bladder contractile activity

Professional appointments

- Medical Faculty, Trakia University, Stara Zagora, Dept of physiology, assistant professor, 1983-1989
- Medical Faculty, Trakia University, Stara Zagora, Dept of physiology, senior assistant professor, 1989-1993
- Medical Faculty, Trakia University, Stara Zagora, Dept of physiology, head assistant professor, 1993
- VAC, 2009, Assoc. Prof.

Membership in Scientific and/or Professional Institutions, Bodies, and Organizations:

- Bulgarian union of scientists
- Bulgarian society of physiological sciences
- Science advisory board
- European peptide society

Publications during Last Three Years:

In foreign journals

1. Zheleva, A., A. Tolekova, M. Zhelev, V. Uzunova, M. Platicanova, v. Gadjeva. Free radical reactions might contribute to severe alpha amanitin hepatotoxicity – A hypothesis, Medical hypotheses, 2007, 69, 361-367. IF-1.276 (2007).
2. Ivanov, I.T., A. Tolekova, P. Chakaarova. Erythrocyte membrane defects in hemolytic anemias found through derivative thermal analysis of electrice impedance. J biochem and biophys methods, 2007, 70(4), 641-648. IF-1.338 (2007)
3. Gadjeva, V., A. Tolekova, M. Vasileva. Effect of spin-labelled 1-ethyl-1-nitrosurea on CCNU-induced oxidative liver injury. Die pharmazie, 2007, 62 (8), 608-13. IF-0.775 (2005)
4. Tolekova, A, G. Ilieva. Hypobaric hypoxia in the regulation of renin-angiotensine sysytem: effects of captopril and nifedipine on plasma renin activity, aldosterone concentration and electrolyte homeostasis. News of medicine and pharmacy, № 274, 59-6.

5. Tolekova A.N., P.V. Hadjibojeva, R.N. Iliev, C.K. Georgiev, K.Y.Trifonova, R.V. Sandeva, R.E. Kalfin, G.S. Ilieva. Participation of extracellular Ca^{2+} or ghrelin in peptide-mediated contraction of strips from rat urinary bladder. Regulatory Peptides, 2010, 162, 77-83, IF 2.39.
6. Yankov, K., A. Tolekova. Influence of Captopril Treatment of Plasma Renin Activity – Mathematical Model, 2010, Notulae Scientia Biologicae, 2(3), 07 - 11.

In Bulgarian journals

1. Ivanova, T., G. Ilieva, A. Tolekova, I. Chakalov. Plasma renin activity after unilateral nephrectomy and nitric oxide synthase inhibition in rats. Trakia journal of sciences, 2007, 5(1) Suppl.1, 26-30.
2. Tsoneva, M., A. Tolekova, V. Gadjeva. Influence of vitamin E on oxidative stress parameters in liver homogenates of mice treated with Cychlomin. Trakia journal of sciences, 2007, 5(1) Suppl.1, 21-25.
3. Antalavicheva, M., P. Goicheva, A. Tolekova, B. Popov, S. PopovaV. Gadjeva. Analysis of plasma concentration of MDA in patients with and without diabetic nephrophaty as manifestation of oxidative stress. Trakia journal of sciences, 2007, 5(1) Suppl.1, 79-82.
4. Tolekova, A., G. Ilieva, B. Popov, S Popova, V. Gadjeva, T. Mircheva. Influence of therapy with recombinant erythropoetin on the oxidative stress of patients with chronic kidney insufficiency. Medical Review, 2007, 43 (4), 55-59.
5. Markova, P., A. Tolekova, G. Ilieva, R. Gurchev. Role of endogenous endotelins in the regulation of plasma renin activity by nitric oxide and renal nerves in spontaneously hypertensive rats. Acta medica bulgarica, 2007, 1
6. Tolekova, A., K. Yankov, Mathematical model of plasma renin activity after nifedipine treatment. Bulgarian Journal of Veterinary Medicine, 2008, 11 (1), 21-29.
7. Ilieva, G., A. Tolekova, R. Sandeva, K. Trifonova, z. Tsokeva, M. Ganeva, Z. Mihova, A. Tolev, S. Zezovski. Role of transmembrane calcium current in angiotensin II – mediated contraction of detrusor organ strips from rat urinary bladder. Bulgarian Journal of Veterinary Medicine, 2008, 11 (2), 89-94.
8. Popov B., V. Gadjeva, P. Valkanov, A. Tolekova, S. Popova Lipid peroxidation and activity of SOD and CAT in astrocytomas. Medical Review, 2008, 44 (2), 53-57.
9. Tsoneva, M., P. Goicheva, A. Tolekova, V. Gadjeva. Influence of spin-labeled antioxidants on oxidative toxicity of cytostatic therapy and ionizing radiation. Trakia journal of sciences, 2008, 6(2) Suppl.1, 32-36.
10. Zheleva, A, A. Tolekova, P. Goicheva. Determination LD50 of potential antitumor agent SLCNUgly and its effect on antioxidant enzyme defence in vivo. Trakia journal of sciences, 2008, 6(2) Suppl.1, 49-53.
11. Ilieva, G., A. Tolekova, R. Sandeva, K. Trifonova, Z. Tsokeva, M. Ganeva. Influence of ghrelin on angiotensin II – mediated contraction of the urinary bladder's smooth-muscle strips. Trakia journal of sciences, 2008, 6 (2) Suppl.1, 77-81. Available online at: http://www.uni-sz.bg
12. Goicheva, P., A. Tolekova, B. Popov, S. Popova, V. Gadjeva, R. Koicheva, D. Sapundjiev. Analysis of activity of SOD and CAT and plasma MDA concentration in patients with and without diabetic nephrophaty as manifestation of oxidative stress. Trakia journal of sciences, 2008, 6(2) Suppl.2, 188-192.

13. Goicheva, P., A. Tolekova, B. Popov, S. Popova, V. Gadjeva, R. Koicheva, D. Sapundjiev. Analysis of activity of SOD and CAT and plasma MDA concentration in patients with insulin-dependent diabetes mellitus, undergoing haemodialysis with polysulfone and cellulose diacetate membranes. Medical Review, 2008, 44(4), 44-49.
14. Hadjibojeva, P., R. Iliev, A. Tolekova, G. Ilieva, K. Trifonova, R. Sandeva. Z. Tsokeva, R. Kalfin, Effects of some neuropeptides on detrusor strips from rat urinary bladder. Bulgarian Journal of Veterinary Medicine, 2009, 12, Suppl. 1, 67−72.
15. Nikolova G., Y. Karamalakova, P. Hadjibojeva, Ts. Georgiev, A. Tolekova,V. Gadjeva, A. Zheleva. Severe mushroom toxin alpha amanitin causes generation of reactive oxygen species in liver tissues of mice – a comparative study by two different instrumental methods. Trakia journal of sciences, 2010, 8, Suppl.2, 149-154.
16. Karamalakova Y., G. Nikolova, Ts. Georgiev, P. Hadjibojeva, A. Tolekova,V. Gadjeva, A. Zheleva. Generation of lipid radicals in mice liver and kidney tissues after treatment by spin labelled nitrosoureas – an ex vivo EPR spectroscopy study. Trakia journal of sciences, 2010, 8, Suppl.2, 144-148.
17. Ivanov, I., A. Tolekova, P. Chakaarova. Temperature and frequency dependence of erythrocyte membrane impedance in case of microspherocitosis anaemia. 20th Anniversary International Scientific Conference, 3rd-4th June 2010, Stara Zagora, Bulgaria (proceedings), v III, 45-49.

In: Anatomy and Physiology Research
Editors: N. R. Zirena and X. B. Liu
ISBN: 978-1-62942-595-8

Chapter 25

Biographical Sketch - Anatomy and Physiology Research

Nicolas Vuillerme

Affiliation:

PhD
AGIM Laboratory UJF CNRS, La Tronche, France

Address:

AGIM Laboratory, AFIRM Team, Facultй de Mйdecine, Bвtiment Jean Roget, 38706 La Tronche Cedex (France)

Date of Birth:

20/06/1976

Education:

2010 - Habilitation in Medicine
2002 – PhD; Sciences and Technologies for Physical Activities and Sports

Research and Professional Experience:

– Assistant Professor at the Grenoble Joseph Fourier University

- Team leader of the Data Acquisition, Fusion and Networks for Medicine research team of the AGIM Laboratory
- Responsible for the Diploma of Research and Technology in Health Engineering of the Grenoble Institute of Technology (Grenoble INP) and Joseph Fourier University (UJF)
- Coordinator of the "E-health" transversal theme within the French CNRS-INSERM Research Network on "Science, Information and Communication Technologies for Health" (GDR STIC-Santй)

Professional Appointments:

2009: Spine Young Investigator Research Award of The International Society for the Study of the Lumbar Spine, Miami, Florida, USA

Publications Last 3 Years:

Internal Journals Papers

1. Pradels, A., Pradon, D., & Vuillerme, N., Effects of experimentally-induced pain on the plantar soles on centre of foot pressure displacements during unperturbed upright stance. *Clinical Biomechanics* (2010) (accepted).
2. Nafati, G., & Vuillerme, N., Decreasing internal focus of attention improves postural control during quiet standing in young healthy adults. *Research Quarterly for Exercise and Sport* (2010) (accepted).
3. Chenu, O., Payan, Y., Demongeot, J., Cannard, F., Diot, B., & Vuillerme, N., Pressure sores prevention for paraplegic people: effects of visual, auditive and tactile supplementations on overpressures distribution in seated posture. *Applied Bionics and Biomechanics* (2010) (accepted).
4. Cuisinier, R., Olivier, I., Troccaz, J., Vuillerme, N., & Nougier, V., Short-term memory effects of an auditory biofeedback on isometric force control: Is there a differential effect as a function of transition trials *Human Movement Science* (2010) (in press).
5. Isableu, B., Ohlmann, T., Crйmieux, J., Vuillerme, N., Gresty, M.A., & Amblard, B., Individual differences in the ability to identify, select and use appropriate frames of reference for perceptuo-motor control. *Neuroscience* 169 (2010) 1199-1215.
6. Roerdink, M., Hlavackova, P., & Vuillerme, N., Center-of-pressure regularity as a marker for attentional investment in postural control: a comparison between sitting and standing postures. *Human Movement Science* (2010) doi:10.1016/j.humov.2010.04.005 (in press).
7. Vuillerme, N., & Boisgontier, M., Changes in the relative contribution of each leg to the control of quiet two-legged stance following unilateral plantar-flexor muscles fatigue. *European Journal of Applied Physiology*, 110 (2010) 207-13.
8. Boisgontier, M., Vuillerme, N., & Iversen M.D., Superimposed electrical stimulation decreases maximal grip force. *The Journal of Sports Medicine and Physical Fitness*, 50 (2010) 152-158.

9. Pinsault, N., Anxionnaz, M., & Vuillerme, N., Cervical joint position sense in rugby players versus non-rugby players. *Physical Therapy in Sport*, 11 (2010) 66-70.
10. Pinsault, N., & Vuillerme, N., Vestibular and neck somatosensory weighting changes with trunk extensor muscle fatigue during quiet standing. *Experimental Brain Research*, 202 (2010) 253-259.
11. Burdet, C., Vuillerme, N., & Rougier, P. How performing a repetitive one-legged stance modifies two-legged postural control. *Journal of Strength and Conditioning Research* (2010) (in press).
12. Pinsault, N., Bouvier, B., Sarrazin, Y., & Vuillerme, N., Effects of vision and tactile stimulation of the neck on postural control during unperturbed stance and cervical joint position sense in young asymptomatic adults. *Spine*, 35 (2010) 1589-1594.
13. Pinsault, N., & Vuillerme, N., Degradation of cervical joint position sense following muscular fatigue in humans. *Spine*, 35 (2010) 294-297.
14. Chenu, O., Vuillerme, N., Demongeot, J., & Payan, Y. A wireless lingual feedback device to reduce overpressures in seated posture: A feasibility study. *PLoS ONE*, 4 (2009) e7550.
15. Vuillerme, N., & Boisgontier, M., Effectiveness of a tongue-placed electrotactile biofeedback to improve ankle force sense following plantar-flexor muscles fatigue. *Gait & Posture*, 30 (2009) 556-559.
16. Fourrй, B., Isableu, B., Gueguen, M., Berbardin, D., Giraudet, G., Vuillerme, N., Pagano, C., & Amorim, M.A., Effect of asymmetrical loading on subjective vertical. *Neuroscience Letters*, 465 (2009) 230-234.
17. Hlavackova, P., Fristios, J., Cuisinier, R., Pinsault, N., Janura, M., & Vuillerme, N., Effects of mirror feedback on upright stance control in elderly trans-femoral amputees. *Archives of Physical Medicine and Rehabilitation,* 90 (2009) 1960-1963.
18. Vaillant, J., Rouland, A., Martignй, P., Braujou, R., Nissen, M.J., Caillat-Miousse, J.L., Vuillerme, N., Nougier, V., & Juvin, R. Massage and mobilization of the feet and ankles in elderly adults: effect on clinical balance performance. *Manual Therapy*, 14 (2009) 661-664.
19. Vuillerme, N., Sporbert, C., & Pinsault, N., Postural adaptation to unilateral hip muscle fatigue during human bipedal standing. *Gait & Posture*, 30 (2009) 122-125.
20. Demongeot, J., Fouquet, Y., Tayyab, M., & Vuillerme, N., Understanding physiological and degenerative natural vision mechanisms to define contrast and contour operators. *PLoS ONE*, 4 (2009) e6010.
21. Vuillerme, N., & Pinsault, N., Experimental neck muscle pain impairs standing balance in humans. *Experimental Brain Research*, 192 (2009) 723-729.
22. Boisgontier, M., Vuillerme, N., Thomas, D., Pinsault, N., Emprin, M., & Caillat-Miousse, J.L., Effects of neuromuscular electrical stimulation on the range of motion recovery in hand proximal interphalangeal sprain. *Science & Sports*, 24 (2009) 192-195.
23. Pinsault, N., & Vuillerme, N., Test-retest reliability of centre of foot pressure measures to assess postural control during unperturbed stance. *Medical Engineering & Physics,* 31 (2009) 276-286.
24. Vuillerme, N., & Cuisinier, R., Sensory supplementation through tongue electro-tactile stimulation to preserve head stabilization in space in the absence of vision. *Investigative Ophthalmology & Visual Science,* 50 (2009) 476-481.

25. Pinsault, N., & Vuillerme, N., The effect of sportive expertise in Free Fly on cervical joint position sense: A pilot study. *Research in Sports Medicine: An International Journal*, 17 (2009) 28-34.
26. Vuillerme, N., Pinsault, N., Chenu, O., Fleury, A., Payan, Y., & Demongeot, J., A wireless embedded tongue tactile biofeedback system for balance control. *Pervasive and Mobile Computing,* 5 (2009) 268-275.
27. Bernard-Demanze, L., Vuillerme, N., Ferry, M., & Berger, L., Can plantar stimulation improve postural control of persons with superficial plantar sensory deficit? *Aging Clinical and Experimental Research,* 21 (2009) 62-68.
28. Zory, R., Vuillerme, N., Pellegrini, B., Schena, F., & Rouard, A., Effect of fatigue on double pole kinematics in sprint cross-country. *Human Movement Science,* 28 (2009) 85-98.
29. Pinsault, N., Vuillerme N., & Pavan, P., Cervicocephalic relocation test to the neutral head position assessment in bilateral labyrinthine-defective and chronic, nontraumatic neck pain patients. *Archives of Physical Medicine and Rehabilitation,* 89 (2008) 2375-2378.
30. Vuillerme, N., Chenu, O., Pinsault, N., Fleury, A., Payan, Y., & Demongeot, J., Can a plantar pressure-based tongue-placed electrotactile biofeedback improve postural control under altered vestibular and neck proprioceptive conditions? *Neuroscience*, 155 (2008) 291-296.
31. Vuillerme, N., Pinsault, N., Fleury, A., Chenu, O., Demongeot, J., Payan, Y., & Pavan, P., Effectiveness of an electro-tactile vestibular substitution system in improving upright postural control in unilateral vestibular-defective patients. *Gait & Posture*, 28 (2008) 711-715.
32. Vuillerme, N., Bertrand, R., Pinsault, N., Postural effects of the scaled display of visual center of foot pressure feedback under different somatosensory conditions at the foot and the ankle. *Archives of Physical Medicine and Rehabilitation,* 89 (2008) 2034-2036.
33. Pinsault, N., & Vuillerme, N., The effects of scale display of visual feedback on postural control during quiet standing in healthy elderly subjects. *Archives of Physical Medicine and Rehabilitation,* 89 (2008) 1772-1774.
34. Vuillerme, N., Pinsault, N., Chenu, O., Fleury, A., Payan, Y., & Demongeot, J., Postural destabilization induced by trunk extensor muscles fatigue is suppressed by use of a plantar pressure-based electro-tactile biofeedback. *European Journal of Applied Physiology*, 104 (2008) 119-25.
35. Pinsault, N., & Vuillerme, N., Differential postural effects of plantar-flexor muscles fatigue under normal, altered and improved vestibular and neck somatosensory conditions. *Experimental Brain Research*, 191 (2008) 99-107.
36. Vuillerme, N., & Boisgontier, M., Muscle fatigue degrades force sense at the ankle joint. *Gait & Posture,* 28 (2008) 521-524.
37. Vaillant, J., Barthalais, N., & Vuillerme, N., Strapping and postural control: effect of underwrap. *Science & Sports*, 23 (2008) 78-82.
38. Vuillerme, N., & Cuisinier, R., Head position-based electrotactile tongue biofeedback affects postural responses to Achilles tendon vibration in humans. *Experimental Brain Research*, 186 (2008) 503-508

In: Anatomy and Physiology Research
Editors: N. R. Zirena and X. B. Liu

ISBN: 978-1-62942-595-8

Chapter 26

Biographical Sketch - Anatomy and Physiology Research

Jinzeng Yang

Date of Birth:

Education:

1985 – BSc; Animal Sciene; Agricultural University of Hebei (China)
1988 – MSc; Animal Genetics; Huazhong Agricultural University (China)
2000 - Ph.D.; Molecular Biology; University of Alberta (Canada)
2002 – Post doctoral; Transgenic animals; USDA-ARS Animal Biotechnology Lab (USA)

Research and Professional Experience:

.. *Positions and Employment*

1985 - 1988: Research Assistant, Huazhong Agricultural University, China
1988 - 1989: Visiting Scientist, Shanghai Medical University, China
1989 - 1994: Research Scientist of Molecular Genetics, HAU, China
1994 - 2000: Research Assistant in Ph.D. program, University of Alberta, Canada
2000 - 2002: Postdoctoral fellow, USDA-ARS Animal Biotechnology Lab, Beltsville, MD
2002 – 2007: Assistant Professor of Animal Biotechnology,
2007- present: Associate Professor of Animal Biotechnology
Chair of Animal Science Graduate Program,
Dept of Human Nutrition, Food & Animal Sciences, University of Hawaii at Manoa

B. Professional Memberships and Other Scholarly Experiences

2007-Present: Member of The Endocrine Society
2005-present, Editor, Animal - An International Journal of Animal Bioscience
2009-Present, Editor, Journal of Agricultural Biotechnology
2003- Present: Ad Hoc Reviewer, Am J Physiol. J Dairy Sci, Endocrinology, J Animal Sci, Genetics, Int J Biol Sci, Mol Bio Rep, Mol Cell Biochem.
2003-Present: Ad Hoc Reviewer, USDA-NIFA, China NSF, Canada NSERC

C. Teaching Experiences

- Genes and Animal Biology (ANSC446, University of Hawaii)
- DNA and Genetic Analysis (MBBE/ANSC650, University of Hawaii)
- Physiology of Domestic Animals (ANSC450, University of Hawaii)
- Lactation Physiology (University of Alberta)

Awards and Honors:

- The Grand Prize of Best Poster, Beltsville Agricultural Research Center, USDA, 2001
- Visiting fellowship in the Canadian Government Laboratories, 1998
- University of Alberta Ph.D. Scholarship ($18000/year), 1994 and 1995
- The First Prize Scientific Research Award, Ministry of Education, China, 1992

Publications Last 3 Years:

1. Kin KH, Kim YS, Yang J. 2010. The muscle-hypertrophic effects of clenbuterol is additive to the hypertrophic effect of myostatin suppression. *Muscle & Nerve* 42: 1012-1017.
2. Li Z, Kawasumi M, Zhao B, Moisyadi S, Yang J. 2010. Transgenic Over-expression of Growth Differentiation Factor 11 Propeptide in Skeleton Results in Transformation of the Seventh Cervical Vertebra into a Thoracic Vertebra *Molecular Reproduction and Development* 77: 1122-1130.
3. Yang Y, Yang J, Liu R, Li H, Luo X, Yang G. 2010. Accumulation of beta-catenin by lithium chloride in porcine myoblast cultures accelerates cell differentiation. Molecular Biology Reports (in press)
4. Watanabe S, Zhao B, Ako H, Yang J. 2010. Identifications of expressed sequence tags from Pacific threadfin skeletal muscle cDNA library. *Aquaculture Research* 41: 572-578.
5. Wang H, Iwai TJ, Zhao B, Lee CS, Yang J. 2010. Identification of microsatellite DNA markers for Pacific threadfin parentage assignment. *J. World Aquaculture Society,* 41: 640-647.

6. Zhao B, Li EJ, Wall RJ, Yang J. 2009. Coordinated patterns of gene expression for adult muscle build-up in transgenic mice expressing myostatin propeptide. *BMC Genomics*. 10:305-315.
7. Li Z, Zhao B, Kim YS, Hu, CY, Yang J. 2009. Administration of a mutated myostatin propeptide to neonatal mice significantly enhances skeletal muscle growth. *Molecular Reproduction and Development* 77:76-82.
8. Laoong-u-thai Y, Zhao B, Phongdara A, Ako H, Yang J. 2009. Identifications of SUMO-1 cDNA and its expression patterns in Pacific white shrimp Litopeanaeus vannamei. *Int J Biol Sci. 5:205-14.*
9. Li Z, B. Cao, Zhao B, Yang X, Fan MZ, Yang J. 2009. Decreased expression of calpain and calpastatin mRNA during development are highly correlated with muscle protein accumulation in neonatal pigs. *Comp. Biochem. Physiol. A: Mol. Integrat. Physiol.* 152:498-503.
10. Poscablo D and Yang J. 2009. Effects of myostatin propeptide transgene on muscle mitochondrial quantity. Ethnicity & Disease 19: S80-S81
11. Cesar, JR, Zhao B, Yang J. 2008. Analysis of expressed sequence tags from abdominal muscle cDNA library of the pacific white shrimp *Litopenaeus vannamei. Animal.* 2: 1377-83.
12. Suzuki,ST, Zhao B, Yang J. 2008. Enhanced muscle by myostatin propeptide increases adipose tissue adiponectin, PPAR-α and PPAR- γ expressions. *Biochem. Biophys. Res. Commun.* 369: 767-73.
13. Yang J, Ferreira R, DuPonte MW, Fukumoto GK, Zhao B. 2008. Growth performances of F1 Angus Plus calves grazing on pasture in Hawaii's tropical climate. *Tropical Animal Health and Production.* 40: 1438-1444.
14. Wu Z, Li Z, Yang J. 2008. Transient transgene transmission to piglets by intrauterine insemination of spermatozoa incubated with DNA fragments. *Molecular Reproduction and Development* 75: 26-32.
15. Bobbili NK, Kim YS, Dunn MA, Yang J, Ong A. 2008. Effects of maternal immunisation against myostatin on post-natal growth and skeletal muscle mass of offspring in mice. *Food and Agricultural Immunology.* 19: 93-106.
16. Cesar J and Yang J. 2007. Expression patterns of ubiquitin, heat shock protein 70, alpha-actin and β-actin over the molt cycle in the abdominal muscle of marine shrimp Litopenaeus vannamei. *Mol. Reprod. Dev* 74: 554-559.
17. Li EJ, Zhao B, Yang J. 2007. Enhanced myogenesis in adult muscle by transgenic expression of myostatin propeptide. *Ethnicity and Disease* 17: S563-564.
18. Yang J, Zhao B. 2006. Postnatal expression of myostatin propeptide cDNA maintained high muscle growth and normal adipose tissue mass in transgenic mice fed a high-fat diet. *Molecular Reproduction and Development* 73:462-469.
19. Cesar J, Zhao B, Malecha S, Ako H, Yang J. 2006. Morphological and biochemical changes in the muscle of the marine shrimp litopenaeus vannamei during the molt cycle. *Aquaculture* 261:688 -694.
20. Borthakur G, Zhao B, Yang J. 2006. Effects of enhanced muscle growth on fatty acid uptake in the skeletal muscle. *Ethnicity and Disease 16: S53-54.*
21. Zhao B, Wall RJ, Yang J. 2005. Transgenic expression of myostatin propeptide prevents diet-induced obesity and insulin resistance. *Biochemical and Biophysical Research Communications.* 337:248-255.

22. Borthakur G, Zhao B, Yang J. 2006. Effects of enhanced muscle growth on fatty acid uptake in the skeletal muscle. *Ethnicity and Disease 16: S53-54.*
23. Yang J, Zhao B, Baracos VE, Kennelly JJ. 2005. Effects of bovine somatotropin on beta-casein mRNA levels in mammary tissue of lactating cows. *Journal of Dairy Science*. 88: 2806-2812.
24. Yang J, Xiong Y. 2003. Genetic manipulations of dominant muscularity genes for efficient pig production. *The Proceedings of International Conferences on Pig Production* (Beijing). Page 67-71.
25. Yang J, Ratovitski T, Brady JP, Solomon MB, Wells KD, Wall RJ. 2001. Expression of Myostatin Pro Domain Results in Muscular Transgenic Mice. *Molecular Reproduction and Development* 60: 351-361.
26. Yang J, Kennelly JJ and Baracos VE. 2000. Transcription factor Stat5 responses to prolactin, growth hormone and IGF-I in rat and bovine mammary gland explant, *Journal of Animal Science* 78: 3114-25.
27. Yang J, Kennelly JJ and Baracos VE. 2000. Physiological levels of Stat5 protein and activity in bovine mammary gland. *Journal of Animal Science* 78: 3126-34.

Part 2 - Research Summaries on Anatomy and Physiology

In: Anatomy and Physiology Research
Editors: N. R. Zirena and X. B. Liu

ISBN: 978-1-62942-595-8

Chapter 27

Brain Glucosensing: Physiological Implications

Sergio Polakof
Universidade de Vigo, Spain

Research Summary

Glucose is an integral part of whole-body energy homeostasis and is tightly regulated by numerous endocrine, neuronal and behavioural systems, which ensure that glucose levels in the blood are maintained within a narrow physiological range. The body continuously adapts its metabolism to keep blood glucose concentrations at a constant value. Glucose homeostasis in man and most other studied mammals is maintained by feedback designed to keep the blood glucose levels close to a set point characteristic for each species. Key to this homeostatic control is the existence of sensors located in different parts of the body that continuously monitor blood glucose variations. They respond to changes in glycemia by triggering hormonal secretion or activation of the autonomic nervous system to control glucose uptake, utilization or production and also to control energy expenditure and food intake. This book examines brain glucosensing and its effect on the body.

In: Anatomy and Physiology Research
Editors: N. R. Zirena and X. B. Liu
ISBN: 978-1-62942-595-8

Chapter 28

Color: Ontological Status and Epistemic Role

Anna Storozhuk
Institute of Philosophy and Law, Novosibirsk, Russia

Research Summary

In this book the complex impact of color action on the organism is discussed. It is shown that the perception of color depends on the action of irritants on other sensor systems and, vice versa, the action of color may exert exciting or inhibiting influence on the perception of sounds or smells. The facts of mutual influence of the sensor systems force us to consider not only a single observation but the entire experience as a continuous flow of sensations which is formulated as the empirical monism principle. This book proposes the solutions of these problems from the point of view of perception's activity and continuity.

In: Anatomy and Physiology Research
Editors: N. R. Zirena and X. B. Liu

ISBN: 978-1-62942-595-8

Chapter 29

Neuroplasticity in the Auditory Brainstem: From Physiology to the Drug Therapy

Angelo Salami
University of Genoa, Italy

Research Summary

It is now known that the brain is actually capable of changing and developing throughout a lifetime. Our mental capacity is astonishingly large, and our ability to process widely varied information and complex new experiences with relative ease can often be surprising. The brain's ability to act and react in ever-changing ways is known as neuroplasticity. Decades of research have now shown that these changes can profoundly alter the pattern of neuronal activation in response to experience. According to the theory of neuroplasticity, thinking, learning, acting and hearing can change both the brain's physical structure and functional organization according to the peripheral stimuli. This book reviews the latest knowledge on the neuroplasticity in the auditory brainstem: from physiology to the drug therapy.

In: Anatomy and Physiology Research
Editors: N. R. Zirena and X. B. Liu
ISBN: 978-1-62942-595-8

Chapter 30

Thecaloscopy: Cadaver Studies

S. Mourgela[1] and A. Sakellaropoulos[2]
[1] Agios Savvas Anticancer Institute, Athens, Greece
[2] Neon Athineon MD Hospital, Athens, Greece

Research Summary

This book contains anatomical and technical information on aspects of endoscopy of the lumbar subarachnoid space, which was named thecaloscopy. For a surgeon it is important to equip all the knowledge and technological advantages for making diagnosis and treatment of patients. Minimally invasive spinal surgery has been well developed in the last few decades. This book suggests the use of flexible endoscopes to approach the content of the thecal sac in order to scrutinize and cure pathological conditions that originate from this anatomical region and cannot be diagnosed by other means.

In: Anatomy and Physiology Research
Editors: N. R. Zirena and X. B. Liu
ISBN: 978-1-62942-595-8

Chapter 31

Vitamin A and Lung Development

***M. L. Pinto*[1], *P. Rodrigues*[1], *A. Honório*[2], *C. Gonçalves*[2] *and V. Bairos*[2]**
[1] University of Trás-os-Montes e Alto Douro, Vila Real, Portugal
[2] University of Coimbra, Portugal

Research Summary

Lung development is a methodical and coordinated process that culminates in the creation of an organ that contains airways, alveoli, blood vessels, multiple innate and acquired host defense systems, and processes to weaken the harmful effects of oxygen in the environment. By regulating cellular differentiation and branching morphogenesis, vitamin A plays an essential role during lung development. Retinoic acid is required for normal alveolar development and vitamin A administration increases lung's elastic fiber contents at birth. Additionally, several experiments showed that retinoic acid restores alveolar architecture to dexamethasone-treated lungs and maintains alveolar development in lungs treated with an angiogenic inhibitor. This book discusses and presents data on vitamin A and its role in lung development.

In: Anatomy and Physiology Research
Editors: N. R. Zirena and X. B. Liu

ISBN: 978-1-62942-595-8

Chapter 32

The Human Biological Clock: From Genes to Chronotherapy

*Michel A. Hofman**
Netherlands Institute for Neuroscience Amsterdam, the Netherlands

Research Summary

The suprachiasmatic nucleus (SCN) and the pineal gland are critical components of a neural oscillator system implicated in the timing of a wide variety of biological processes. The circadian cycles established by this biological clock occur throughout nature and have a period of approximately 24 hours. With advancing age these daily fluctuations deteriorate, leading to disrupted cycles with a reduced amplitude. In humans, age-related changes have been described for hormonal rhythms, body core temperature, sleep-wakefulness, and several other behavioral cycles. It appears that the disruption of circadian rhythms and the increased incidence of disturbed sleep during aging are paralleled by age-related alterations in the neural and temporal organization of the circadian oscillator and a decreased photic input to the clock. The observed neuronal degeneration of the SCN-pineal complex in senescence strongly suggests an organic deterioration of the circadian oscillator during aging and supports the idea that damage to the biological clock is the underlying anatomical substrate for the clinically often-observed disturbances in circadian rhythmicity in mood disorders and Alzheimer's disease. Recent studies furthermore indicate that illuminance and photoperiod may be considered as potential environmental factors controlling the functional activity of the circadian timing system. If light is the effective Zeitgeber in man, as it is in many other organisms, disturbances in processing of light information during aging or in some neurodegenerative and affective diseases may have profound effects on the timing of a variety of physiological and behavioral activities, including sleep. In view of the assumed sensitivity of the biological clock to changes in day length and light intensity research is now focussing

* Correspondence: M.A. Hofman. Ph.D. Netherlands Institute for Neurosciene, Meibergdreef 47, 1105 BA Amsterdam. The Netherlands. Telephone number: +31-20-5665500. Fax number: +31-20-6961006. E-mail: M.Hofman@nin.knaw.nl.

on whether the synchronizing and antidepressant effects of bright light exposure in sleep disturbances and mood disorders can be improved by applying the therapy in temporal coherence with the circadian cycles underlying the neuronal activity of the biological clock.

In: Anatomy and Physiology Research
Editors: N. R. Zirena and X. B. Liu
ISBN: 978-1-62942-595-8

Chapter 33

Photosensitivity A Disregarded Attribute to Analyze "Photoperiodic Clocks"

Hubert R. Spieth* and Katharina Strauß
University of Bielefeld, Department of Animal Ecology, Bielefeld, Germany

Research Summary

To interpret seasonal timing in arthropods the change of photosensitivity to light/dark cycles during certain ontogenetic phases has been more or less ignored when modelling photoperiodic responses. The consequences of changing photosensitivity are exemplified using the large white butterfly, *Pieris brassicae*, and its response to daylength. It is an essential character in two ways: 1) Because of the 'required day number' (RDN) necessary to induce a specific response, which changes with photosensitivity. 2) Because of perception of increasing and decreasing daylength, which is not a specific ability of a species but a consequence of changing photosensitivity.

With photosensitivity-tests, evidence for a general concept of quantitative time measurement in arthropods is presented. In this context the findings reveal a converse effect of the same photoperiod depending on the light/dark regime in which an individual grows up. This opposite effect supports the argument for the existence of two independent targets for light/dark-cycles, interpreted as two antagonistic time measurement systems, and gives evidence for a 'double circadian oscillator clock' mechanism which is based on two submechanisms, a 'short-night determining system' and a separate 'long-night determining system'. The existence and independence of two systems is shown by differences in long-night vs. short-night responses regarding photosensitivity, temperature dependence, duration of diapause, and heritable factors.

* Correspondence Author: Phone: +49 521 1062701 E-mail : hubert.spieth@uni-bielefeld.de.

In: Anatomy and Physiology Research
Editors: N. R. Zirena and X. B. Liu

ISBN: 978-1-62942-595-8

Chapter 34

Respiration Rhythmic and Quality of Sleep to the Total Wellness and Development of a Child

***Jong Yong Abdiel Foo*[1,2*] and *Stephen James Wilson*[3]**
[1]Division of Research, Singapore General Hospital, Singapore
[2]Bioprocessing Technology Institute,
Agency for Science Technology and Research, Singapore
[3]School of Information Technology and Electrical Engineering,
University of Queensland, Australia

Research Summary

There is an increasing interest to improve the wellbeing of life especially amongst the middle to upper social class in the well-developed nations. Particularly, the study on quality sleep each night has refueled much attention to the overall wellness of these individuals. Studies have shown that disturbances during sleep not only can lead to mild irritations but also adverse events and life satisfaction as a whole. At the heart of the issue, the rhythmical functions of the body can be distorted when the occurrences of sleep disturbances become frequent. Sleep is now no longer viewed simply as a state of rest but it has specific and affirmative effects. Essentially, the brain is the prime beneficiary of sleep in order to maintain its cerebral capacities. Hormonal changes under the control of the sympathetic nervous system can increase blood flow to the muscles. This promotes the breakdown of stored nutrients into glucose in muscle tissue, thus providing energy available for the body to use. Monitoring the quality of sleep has stretched beyond just for individuals with susceptible sleep disorders but for many who want to maximize their daily lives. Compound with the fact that in many well-developed countries, birth rate has been low; caregivers are becoming more concerned in bestowing prime care for the long-term development and wellness of their children. However, limited research

* Correspondence Author: Jong Y.A. Foo. Bioprocessing Technology Institute, Agency for Science Technology and Research (A*STAR), 20 Biopolis Way, #06-01 Centros, Singapore 138668. Email: abdielfoo@hotmail.com.

has been initiated to assess the quality of sleep in children until the recent decades. This can be due to the misconception or poor understanding of caregivers in pediatric sleep disorders. With much recent public knowledge in this domain, there is a paradigm shift in the mindset of caregivers. Thus, the once much-neglected yet specialized field of study has evolved since, from complex studies only conducted in fully equipped sleep laboratories to possibly simplified, miniature and portable ambulatory based studies. In this chapter, the comprehensive coverage of this topic will be presented together with a general perspective from a child, caregiver and healthcare provider.

In: Anatomy and Physiology Research
Editors: N. R. Zirena and X. B. Liu

ISBN: 978-1-62942-595-8

Chapter 35

Circadian Sleep-Wake Rhythms in Preterm Infants

Ronny Geva and Ruth Feldman
The Gonda Multidisciplinary Brain Research Center
Bar-Ilan University, Israel

Research Summary

Knowledge is recently building regarding the factors that affect the development of an optimal sleep-wake cycle. Fetuses near term experience arousal fluctuations and sleep episodes already in the darkness of the womb. Yet, the factors that allow for a smooth emergence of an optimal sleep wake cycle are the conditions that place newborn at higher risk to develop sleep disorders are not yet fully understood. The current chapter proposes an integrative model for the development of circadian sleep-wake rhythms in preterm infants. This model is based on a comprehensive review of the basic science literature as well as clinical work with at-risk preterm infants. This integrative model proposed a three source risk source: 1) infant dependent neurobiological vulnerability risk, such as brainstem mediated functional disability; 2) a familial source that entails genetic predispositions, parental style and support resources; and 3) an environmental stimulation source, such as excessive NICU related stimulation and handling. These three realms act as main effects and interact with the infants' prenatal and postnatal age and with the infants' self- regulation mechanisms. The manner in which these sources operate and interact lead to directions of interventions that ameliorate the development of circadian sleep-wake rhythm deficit in infants born at-risk for sleep disorders. The chapter details each component of the model and illuminates the complexity of the interactions involved in it to deepen the understanding the mechanisms involved in early organization of sleep-wake rhythms in infants.

In: Anatomy and Physiology Research
Editors: N. R. Zirena and X. B. Liu
ISBN: 978-1-62942-595-8

Chapter 36

Interactions between the Circadian and the Immune System: A Framework for the Understanding of Disease

Natalia Paladino, María Juliana Leone, Leandro P. Casiraghi, Patricia V. Agostino, Diego A. Golombek and Juan J. Chiesa*
Chronobiology Laboratory, National University of Quilmes/CONICET, Buenos Aires, Argentina

Research Summary

The daily environmental changes have imposed a selective pressure for life on Earth, driving the development of a circadian clock mechanism for the generation and entrainment of rhythms in physiological and behavioral variables (e.g. body temperature, hormonal secretion, sleep, locomotor activity, etc.). In mammals, the clock resides in the hypothalamic suprachiasmatic nuclei (SCN), and the principal signal that adjusts its activity is the light-dark cycle. Most immune factors and processes are under circadian control, although the efferent pathways that control these cycles are not completely understood. In addition, circadian disorders are usually associated with disease, and temporal disarrangements in immune parameters are closely related to the onset or the development of pathologies. In particular, several lines of evidence link the circadian clock to cancer disease. Epidemiologic studies show associations between circadian rhythms disruption and cancer risk or inadequate responses to treatment. Moreover, proinflammatory cytokines, some of which have been reported to be key actors in tumour signaling, can generate deep behavioral alterations comprising clock-associated variables during tumor progression, such as altered sleep pattern, fatigue, and lower quality of life.

---

*Correspondence Author: Dr. Diego Golombek, Departamento de Ciencia y Tecnología, Universidad Nacional de Quilmes, R.S. Peña 352, (1876) Bernal, Buenos Aires, Argentina. Fax +54-11-4365-7132. E-mail: dgolombek@unq.edu.ar.

We argue that the circadian-immune interaction operates in two directions: while the SCN (or other, peripheral, clocks) might drive circadian variations in several immune variables, humoral signals can in turn affect the molecular mechanism of the circadian clock and its entrainment. For example, higher diurnal lethality depicts a strong circadian modulation of lipopolysaccharide (LPS)-induced sepsis, which is also observed using TNF-α stimulation. Current clinical data show a strong correlation between time of day and illness manifestation or immune activity. In the other direction, administration of subpyrogenic doses of LPS, as well as of proinflammatory cytokines IL-1β and TNF-α induces changes in SCN-driven locomotor activity rhythms. At the molecular level, this effect is achieved through activation of a transcriptional pathway involving nuclear factor-κB (NF-κB). This and other pathways (such as mitogen-activated-kinase (MAPK) signaling), lead to the expression of specific genes of the circadian clockwork, which in turns regulates its rhythmic output. Regarding cancer disease, the clock genes *Per1* and *Per2* are involved in DNA damage response, and could therefore regulate tumour suppression. Moreover, the expression of many cell-cycle genes is under circadian control. Alterations in other components of the molecular clock, as those in members of the casein kinase (CK) family, have also been linked to cancer disease.

In this review we analyze the interaction between the circadian and immune systems, and explore the signal transduction pathways which could participate in this dialogue. The precise knowledge of this interaction might be extremely useful for the understanding of diseases development such as cancer.

In: Anatomy and Physiology Research
Editors: N. R. Zirena and X. B. Liu
ISBN: 978-1-62942-595-8

Chapter 37

Chronoecology of Neotropical Primates: The Spider Monkey Ateles Geoffroyi

Jairo Muñoz-Delgado[1] and María Corsi-Cabrera[2]
[1]Chronoecology and Human Ethology Research-Group.
Direction of Neurosciences.
National Institute of Psychiatry Ramón de la Fuente Muñiz,
Mexico City, and Faculty of Psychology, UNAM, Mexico
[2]Sleep Laboratory. Faculty of Psychology,
National Autonomous University of Mexico, UNAM, Mexico

Research Summary

The emerging discipline denominated "chronoecology of behavior" is the result of the interaction between the work of behavioral ecologists –those who study behavioral phenomena in the natural environment– and chronobiologists, who explain biological rhythms on the basis of experimental studies carried out in the laboratory. Since the beginning of this century, this interaction has generated interest in understanding how biological rhythms behave in organisms living in their natural habitat. This approach requires field studies of rest/activity rhythms in order to obtain detailed hour/hour and day/day information about the possible effects of the modulation of certain environmental variables, such as photoperiod, temperature and the lunar sky, which may synchronize and/or mask behavioral activity. As a model for the study, a neotropical monkey species was chosen, the spider monkey, which is characterized by being widespread in America, from Mexico to southern Bolivia, and by its fusion-fission society and vast behavioral and vocal repertoire. In carrying out the studies, motor activity was registered using *Actiwatch* accelerometers during 24 continuous hours for 12 months, in combination with focal recordings of behavior and geophysical variables. On the basis of the results, the role of geophysical (astronomical and meteorological) variations as *zeitgeber* and/or the masking of the circadian rhythm, contributed valuable information for understanding alterations in physiology and behavior introduced by temporal adaptations to environmental changes. Finally, it is well-known that behavioral variations in non-human primates and *Homo sapiens* are influenced by astronomical and meteorological factors and by social, vocal, verbal and non-verbal conspecific interactions; therefore, it is essential to dissect these variations in order to study them and find out how the circadian rhythm modulates behavior and in what way social interaction influences the rhythms of circadian activity.

In: Anatomy and Physiology Research
Editors: N. R. Zirena and X. B. Liu
ISBN: 978-1-62942-595-8

Chapter 38

On Clocks, Chaos and Cancer: A Biodynamic Approach to Cancer

Federico Cardona[*]
Wiedemann Parkklinik, Biomedical Research Dept., Markdorf, Germany

Research Summary

Self-organizing of open dynamic systems far from thermodynamic equilibrium gives rise to complexity in space and time, thus creating new structures including biological clocks. In fact, periodic oscillations can be used to control deterministic chaos. The human organism is a multioscillator, which is governed by the master clock in the hypothalamus. We have previously observed a periodic oscillation of free radical production that functions synchronous to the central clock and is apparently generated by it. We found as well, that this oscillation, possibly mediated by mitochondrial activity, ceases to function in humans suffering from advanced cancer diseases. Alterations of mitochondrial bioenergetics have been observed to be a general feature of malignant cells and could progressively diminish the distance from thermodynamic equilibrium necessary to maintain multicellular complexity. Self-organization could thus break down, leading to malignant development as a survival strategy of tumor cells.

[*] Corresponding Author: fcardona@web.de.

In: Anatomy and Physiology Research
Editors: N. R. Zirena and X. B. Liu
ISBN: 978-1-62942-595-8

Chapter 39

Klepsydraic Model of Internal Time Representation: Experimental Findings and Analytical Properties

Jiří Wackermann
Dept. of Empirical and Analytical Psychophysics, Institute for Frontier Areas of Psychology and Mental Health, Freiburg i.Br., Germany

Research Summary

In this chapter a "klepsydraic" model of internal time representation is introduced. The model's working is illustrated by experimental data on duration reproduction and duration discrimination. In addition, special properties of the reproduction function predicted by the model are discussed. It is shown that time-scales generated by "klepsydraic clocks," although non-uniform with respect to the objective (physical) time, allow for rational time-keeping. These findings recommend the model as an analytic instrument in studies of time perception, timing behavior and subjective time awareness.

In: Anatomy and Physiology Research
Editors: N. R. Zirena and X. B. Liu
ISBN: 978-1-62942-595-8

Chapter 40

Cortical and Subcortical Processing of Color: A Dual Processing Model of Visual Inputs

Hitoshi Sasaki
Department of Physiology and Biosignaling, Osaka University
Graduate School of Medicine, Yamadaoka, Japan

Research Summary

In order to investigate a possible role of visual processing in regulation of adaptive behaviors, two behavioral experiments using color stimulus were performed. In the first experiment, hemispheric asymmetry of color processing was investigated by measuring reaction time to a stimulus presented either in the left or the right visual field responded by ipsilateral hand. The simple reaction time was shorter to a color stimulus presented in the right hemisphere in the right-handed participants, while no hemispheric asymmetry was found in color discrimination reaction time without verbal cues. In the second experiment, a modulatory effect of color on sensory-motor interaction was investigated using a prepulse modulation task. Amplitude of a startle eyeblink response elicited by an air-puff to the cornea was significantly inhibited by a shortly (100 ms) preceding color prepulse. A different amount of the inhibition was induced by different color prepulses. Yellow was more effective as compared to a blue prepulse. Although the exact neuronal pathways underlying the prepulse inhibition of the corneal blink response is not determined, a top-down pathway from the cortex to the brainstem nuclei via the amygdala seems to be involved in the sensory-motor interaction. The descending pathway seems to play a role in modulation of the startle responses. From these findings combined with other studies, a dual processing hypothesis of visual inputs will be proposed, where physical features of the stimulus are processed in the cerebral cortex with consciousness, while the psychological and biological meanings are processed mainly in the limbic system without consciousness. Traditionally, it was thought that these two processes are in series, while in the present model these processes are in parallel, in addition to the serial processing. Visual inputs are conveyed to the limbic system via the indirect cortical and the direct subcortical pathways. The cortical pathway further divided into two routs; one is from the inferotemporal cortex and the other is from the posterior parietal association cortex through the pulvinar nucleus of the thalamus.

In: Anatomy and Physiology Research
Editors: N. R. Zirena and X. B. Liu

ISBN: 978-1-62942-595-8

Chapter 41

Color: Ontological Status and Epistemic Role

Anna Storozhuk
Fellow in Philosophy at the Institute of Philosophy and Law, Novosibirsk, Russia.
Tutor at Novosibirsk State University, Russia

Research Summary

There are two basic approaches to studying color: one of them considers the issue of the physical reasons of color, the other investigates color perception. According to the first, color is not an objective physical entity; the second approach has many experimental evidences of color influence on human organism, for example, changes of the emotional condition, blood pressure, accuracy of perception, etc. The considered question can be formulated as follows: how color, being a sign without a referent, can make a real impact on the organism? A working hypothesis is that color is a non-conventional objective sign. This hypothesis will be subjected to critical analysis from the point of view of psychology of development in order to ascertain whether the sign properties of color are innate or are formed by the influence of culture. Another topic is a role of color in the world cognition. This question was usually considered from the point of view of direct influence on the increase of the visual recognition accuracy. We will investigate a question of indirect influence of color by means of pre-setting of the nervous system to perception; this is possible thanks to the system character of perception.

In: Anatomy and Physiology Research
Editors: N. R. Zirena and X. B. Liu
ISBN: 978-1-62942-595-8

Chapter 42

The Biological Significance of Colour Perception

Birgitta Dresp-Langley and Keith Langley*
Universite de Montpellier, Montpellier, France

Research Summary

There is no colour without light, nor is there colour perception without a sensory organ and a brain to process visual input. This chapter first reviews how the colour of objects is produced. Most commonly, this depends on so-called pigments, the molecular nature of which provokes strong absorption of part of the incident light falling on an object. Colour can also be produced by optical phenomena such as refraction, dispersion, interference or diffraction from ordered structures within objects. A wide variety of photonic microstructures are known in the living world and specific examples will be described in mammals, birds, fish and insects. Some of these structures reflect light in the near ultraviolet spectral region, particularly pertinent for certain birds, insects and fish which are sensitive to these wavelengths. A detailed account of a particularly elaborate structure present in the king penguin beak will be given to illustrate the extent to which evolutionary pressure leads to the elaboration of such structures to satisfy specific needs of birds or animals. Subsequently, the perception of colour in man and animals and its biological significance is dealt with. For man, this will include a discussion of the symbolic meaning of different colours. In many species, especially birds, the colours of plumage and parts of the skin have an important survival function. Such biological colourations may fulfil the role of ornaments that determine mate choice and reproduction of the species, or signal good health, allowing individuals to secure and maintain territorial dominance. Colour perception may also have an underlying survival function in man, but more complex explanations are needed to relate perception to such a function. The colour of an object in the visual field is known to determine the way in which humans perceive relations between objects and their background, particularly which objects

* e-mail: klangley@numericable.fr.

appear nearer. This suggests that colour perception is important in processing information about the physical structure of the world. The colour red plays an important role in this process, since it drives mechanisms of visual selection which attract attention to, or away from, objects in the visual field. Psychophysical studies of colour perception in both animals and man help to understand these complex processes. Finally, colour perception in man may contribute either to rewarding psychological sensations of warmth, comfort and safety or to aversive sensations of coldness and discomfort, sensations which can strongly influence individuals in their daily social interactions.

In: Anatomy and Physiology Research
Editors: N. R. Zirena and X. B. Liu
ISBN: 978-1-62942-595-8

Chapter 43

Individual Differences in Colour Vision

M. I. Suero, P. J. Pardo and A. L. Pérez
Dept. of Physics, Avda. de Elvas s/n, Badajoz, University of Extremadura, Spain

Research Summary

Since the formulation of the Young-Helmholtz chromatic theory in the 19th century, it has generally been accepted that human colour vision is trivariant, i.e., it is possible to match any colour stimulus by mixing three primary stimuli in appropriate proportions. This resulted in the definition by the International Commission on Illumination in 1931 of the standard colorimetric observer for fields of 2° only three years after the definition of the standard photometric observer known as the V_λ luminous efficiency curve.Much progress has been made in the knowledge of colour vision since then, in such fields as physics, physiology, genetics, biochemistry, neuroscience, and psychology. That is why it is perhaps the time to raise our level of exigency a step when it comes to characterizing a chromatic observer. The question we should ask ourselves is not whether that figure of the standard colorimetric observer represents the average of the population, because this can indeed be done with successive corrections. Rather it is whether there exist very large individual differences in the perception of colour, and if so how these differences manifest themselves. This is of course apart from observers characterized as defective.In this chapter, we review the state-of-the-art in this field, and present our own latest research results concerning this question.

In: Anatomy and Physiology Research
Editors: N. R. Zirena and X. B. Liu
ISBN: 978-1-62942-595-8

Chapter 44

Color-Sensitive Neurons in the Visual Cortex: An Interactive View of the Visual System

Maria C. Romero,[1] Ana F. Vicente,[1] Maria A. Bermudez[1] and Francisco Gonzalez[1,2]
[1]Department of Physiology, School of Medicine, University of Santiago de Compostela, Santiago de Compostela, Spain
[2]Service of Ophthalmology, Complejo Hospitalario Universitario de Santiago de Compostela, Santiago de Compostela, Spain

Research Summary

Classically, different physical attributes of the visual stimulus were thought to be solved in parallel by interdependent neuronal populations conveying information from the retina to the parietal and temporal cortical areas. According to this assumption, while neurons in the dorsal areas of the visual system were mainly related to the analysis of motion and spatial information, those located at the more ventral positions were mostly associated to shape and color processing. However, although this functional segregation between visual areas has been supported for several decades, there is also strong experimental evidence suggesting an alternative task-driven view of the visual system. According to this more recent perspective, neuronal responses in cortical visual areas can be simultaneously dependent on more than one single visual attribute. As far as color perception plays a central role in visual recognition, it could be assumed that color-sensitive neurons would be also involved in the analysis of some other critical visual attributes. In agreement with this idea, it has been shown that V1 double opponent cells respond to edges defined not only by chromatic and luminance differences, but also by the orientation of their receptive fields. Furthermore, results from many electrophysiological and neuroimaging studies have also demonstrated that color-sensitive neurons in V2 and V3, modulate their responses depending on diverse physical attributes of the stimulus such as the stimulus direction, orientation, luminance and shape, revealing the simultaneous contribution of magno- and parvocellular inputs from the Lateral Geniculate Nucleus (LGN) at different levels of the visual system. At higher visual areas, several authors have reported the existence of multi-sensitive neurons. Middle Temporal (MT) neurons, in the dorsal stream, are sensitive to motion spots defined by single or combined changes in texture and color. In the ventral stream, responses to both, color and orientation have been described in V4 and the inferotemporal cortex. Additionally, results from several studies blocking the magno- and parvocellular projections from the LGN to V4 have shown that these two channels can simultaneously contribute to neuronal responses at this level of processing. All these data evidence that even sharply-color-tuned neurons can show color-related responses modulated by many other visual attributes.

In: Anatomy and Physiology Research
Editors: N. R. Zirena and X. B. Liu
ISBN: 978-1-62942-595-8

Chapter 45

Is Colour Composition Phenomenal?*

Vivian Mizrahi
Department of Philosophy, University of Fribourg, Switzerland

Research Summary

Colour composition divides colours into two types: unitary and binary colours. Colours which are not composed are said to be "unique" or "unitary" colours, whereas composed colours are always binary. Colour composition and the distinction between unitary and binary colours have played a major role in colour science and in the philosophy of colours. They have for example been invoked to introduce opponent-processes in the mechanisms underlying colour vision and have been used to criticize philosophers who defend a physicalist view on the nature of colours. Most philosophical or scientific theories suppose that colour composition judgments refer to the way colours appear to us. The dominant view is therefore *phenomenalist* in the sense that colour composition is *phenomenally* given to perceivers. This paper argues that there is no evidence for a phenomenalist view of colour composition and that a conventionalist approach should be favoured.

* Thanks to M. Nida-Rümelin, J. Suarez, Y. Jraissati and D. Gross for helpful comments and criticisms. This paper was written with the support of a grant from the SNF (Swiss National Science Foundation).

In: Anatomy and Physiology Research
Editors: N. R. Zirena and X. B. Liu

ISBN: 978-1-62942-595-8

Chapter 46

Color Image Restoration and the Application to Color Photo Denoising

*Lei He**
Department of Information Technology, Armstrong Atlantic State University, Savannah, GA, US

Research Summary

Image restoration has been a classical and significant topic of image processing, which refers to the techniques to reconstruct or recover an image from distortion (e.g. motion blur and noise) in different applications, such as satellite imaging, medical imaging, astronomical imaging, and family portraits. For motion blur, *image deblurring* techniques are used to estimate the actual blurring function and "undo" the blur to restore the original image. In cases where the image is corrupted by noise, *image denoising* methods are employed to compensate for the degradation the noise caused. In the past two decades, image denoising has been a fundamental and active research topic and widely used as a key step in a variety of image processing and computer vision applications, such as image segmentation, compression, object recognition, and tracking. This chapter focuses on image denoising, specifically for color image denoising and the application to color photo denoising.

Color image denoising has been an active area with the fast progress of optical camera techniques in past decades. There are several major color spaces to represent color images and the most common one is the Red, Green and Blue (RGB) model. Other color systems include YIQ (luminance, hue and saturation) system (NTSC), YCbCr (luminance, blue minus luminance and red minus luminance) system, CMY and CMYK (cyan, magenta, yellow and black) systems, HSV (hue, saturation and value) and HSI (hue, saturation and intensity) systems. The selection of a system is generally application-specific, e.g. RGB for education and presentation, NTSC for television, YCbCr for digital video, HSV for color palette.

* Lei.He@armstrong.edu.

Among these color spaces, HSI space has been well accepted for its favorable color description capability that is practical for human interpretation.

As a specific application of color image denoising, color photo noise removal is an important subject due to extensive use of digital cameras in recent years. Color photo noise is introduced in the process of image acquisition, filtering, compression and reconstruction. In digital photography, a high ISO setting of a digital camera is usually used to increase light sensitivity in dark environments, but the resultant image contains more sensor noise than a low ISO image taken with the same exposure*. The objective of our work is to denoise high ISO photos to achieve a low ISO photo quality without the use of expensive cameras or accessories. The motivation comes from the practical needs of digital camera users and manufacturers. So far, most existing approaches are used to solve a general denoising problem (e.g. gray, color or multi-spectral images) based on an assumption of additive or multiplicative noise independent of signal and to the best of our knowledge, none of presented literatures focus on the unknown digital camera sensor noise in color photos.

Since the pioneering work in regularization methods, scale-based analysis has played an increasingly important role in signal (image) processing. The idea is to represent an image in multi-scale, so that only important features are preserved and unwanted features (e.g. noise) are removed in low scales. This is based on the conclusion that convolving a signal with a Gaussian kernel is equivalent to evolving it with a heat differential operator where time is the scale. Such methods have direct applications for feature extraction and image denoising. For example, given an image I: $\Omega \subset \Re^2 \rightarrow \Re^m$, the corresponding heat diffusion equation is:

$$\frac{\partial I}{\partial t} = I_t = \mathbf{div}(\nabla I) = I_{\eta\eta} + I_{\xi\xi}, \tag{1}$$

where η and ξ refer to the normal and tangential directions respectively. For the denoising applications, this linear filtering approach corresponds to isotropic diffusion, which presents a major limitation: important details in an image also get smoothed away along with the noise. Therefore, a nonlinear filter or an anisotropic diffusion is needed to preserve those details from the smoothing process. Since the seminal model proposed by Perona and Malik in the early 1990s, numerous literatures have been presented to recover the "true" image from noisy data through a nonlinear analysis, such as partial differential equation (PDE)-based anisotropic diffusions, variational approaches, robust statistics, as well as some transformation based approaches (e.g. wavelet, ridgelet and curvelet-based image denoising), just to name a few. A complete review of current image denoising approaches can also be seen in several recent literatures.

This chapter presents a novel framework specifically for denoising color photos particularly those photos taken at a high ISO setting that resulted in noticeable sensor noise. Compared with existing literatures, there are two major contributions in the presented approach. The first is that the color photo denoising is conducted in the HSI space instead of the traditional RGB space, which is motivated by the fact that the HSI model has a better color description for human interpretation. Our algorithm is based on separating a color photo into hue, saturation, and intensity components, and then processing each component with

* This is achieved by reducing the shutter speed or increasing the aperture.

PDEs or diffusion flows. The intensity denoising is our main focus, which is implemented with a PDE that is a combination of a gradient vector flow (GVF)-based filter and a fourth-order PDE filter. This combined technique provides a robust and accurate denoising process, i.e., it preserves edges well and at the same time overcomes the staircase effect in smooth regions. The hue and saturation denoising are implemented by a weighted orientation diffusion and a modified curvature diffusion respectively. The second contribution is the algorithm performance assessment. In most existing literatures, synthetic noise (e.g. Gaussian, salt & pepper noise) has been used for algorithm performance evaluation. Furthermore, the denoising effect has been assessed mostly by human visual perception, few by the mean squared error (MSE) and the peak signal to noise ratio (PSNR). In contrast, the color photo noise produced by digital cameras is real sensor noise. In this chapter, the proposed algorithm is evaluated by comparing the denoised images with the ground truth. The ground truth and noisy images are produced by the same camera on the same scene with low and high ISO settings by maintaining the same exposure. Because ground truth images are available for those photos taken under a controlled environment, we could assess the performance of our proposed algorithm, several recognized methods, and commercial software using objective error measurements of MSE and PSNR, in addition to the commonly used subjective visual assessment. Both qualitative and quantitative validation shows that the proposed algorithm is more appropriate for color photo denoising than existing approaches.

In: Anatomy and Physiology Research
Editors: N. R. Zirena and X. B. Liu

ISBN: 978-1-62942-595-8

Chapter 47

Color in Psychological Research: Toward a Systematic Method of Measurement

Roger Feltman and Andrew Elliot
University of Rochester, NY, US

Research Summary

Psychology is a discipline that prides itself on being an empirical science. As such, rigorous statistical and methodological controls must be used to ensure the validity of every result. Ostensibly, each submission for publication is peer reviewed, and needs to be replicated by other scientists in other locations to confirm or disconfirm the results. This is how a scientific discipline must operate if it wishes to produce meaningful, accurate results. When a discipline strays from these procedures, it leaves itself open to criticism and more importantly, to the possibility of inaccurate or misleading conclusions. All research needs to ascribe to these standards, regardless of how time consuming, inefficient, or difficult they may be.

One area of research that has failed to live up to these standards is the study of color and psychological functioning. The aesthetic property of color may at first consideration make it seem like a trivial topic for study, but recent research indicates exactly the opposite. Color has been shown to influence affect, cognition, and behavior. The degree and type of influence has varied from study to study, some more psychologically consequential (e.g. color and performance: Elliot et al., 2007) than others (e.g. shoe color preference: Trinkaus, 1991). None of these results, however, can be considered valid if they fail to live up the methodological rigors of science.

An in depth examination of the color research of the past and present makes it clear that most of the work fails to meet scientific standards. Too many studies have failed to take into account the three basic properties of color. Others have failed to consider the unconscious associations people have with color that could act as possible confounds. Stated differently, color used in an experiment may affect the experiment's dependent variable in unwanted and unaccounted for ways. In either case, it is impossible to draw meaningful conclusions from these studies, as their results could be due to any number of variables. This is the primary argument that will be made throughout this chapter. The aim is not to criticize or demean the existing research or researchers. Rather, it is hoped that this analysis will lead to more systematic, scientifically valid empirical work on color psychology. By learning about and avoiding the mistakes documented below, researchers will be able to meaningfully add to the growing body of work in on color psychology.

In: Anatomy and Physiology Research
Editors: N. R. Zirena and X. B. Liu
ISBN: 978-1-62942-595-8

Chapter 48

Color in Aquaculture. An Importance of Carotenoids Pigments in Aquaculture of Salmon and Echinoderms

***Pavel A. Zadorozhny*[2], *Marianna V. Kalinina*[1], *Eugene V. Yakush*[1] *and Eugene E. Borisovets*[1]**
[1] Pacific Fisheries Research Centre (TINRO), Vladivostok, Russia
[2] Institute of Chemistry, Far Eastern Branch of Russian Academy of Science, Vladivostok, Russia

Research Summary

Carotenoids are widely used in aquaculture to achieve natural coloring of salmon flesh, improvement of trade quality (color) of sea urchin roe and in aquaculture of Crustaceans.For salmon, it has been found, that the relationship between pigment content and color parameters is complex and nonlinear. Nevertheless, there is an evident correlation between the total concentration of carotenoids (mainly astaxanthin) and the red, most valued by consumers, color of a muscular tissue of salmon (i.e. the higher the pigment content the better). Assimilation of carotenoids in salmon usually does not exceed 10-15 per cent, and cost of astaxanthin makes up about 6-8 % from the cost of filleted fish. Thus, researches in this field are directed on improvement of feed composition increasing of carotenoid assimilation and search of new sources of these pigments; optimization of processing and storage conditions of production, allowing keeping natural color. Ability to reach desirable color of roe is crucial condition for commercial echinoculture. A number of studies were devoted to developing of composition of artificial feed giving desirable color characteristics. Considering macroalgae, the best results have been reached with species of Laminaria, Alaria, Palmaria, and Ulva. It has been proved great significance of carotenoids as essential micronutrients for sea urchin aquaculture. A promising source of carotenoids in aquaculture may be microalga Duneliella salina. Carotenoid content correlates with redness of the gonads, but unlike salmon, for sea urchins there is a certain optimum of the pigment concentrations in gonads, excess or, on the contrary, lack of the pigments lead to falling into less desirable for customers color grades.

In: Anatomy and Physiology Research
Editors: N. R. Zirena and X. B. Liu
ISBN: 978-1-62942-595-8

Chapter 49

Black Enough?: Needed Examination of Skin Color among Corporate America

Matthew S. Harrison and Wendy Reynolds-Dobbs
The University of Georgia, US

Research Summary

A common problem among social scientists who group all members of a race/ethnicity together is that they assume that all of the life experiences of those individuals are the same, and thereby, overlook the prevalence of heterogeneity within ethnicities. One such example is a global phenomenon present in all cultures where there is skin tone variation—colorism. This longstanding ideology which suggests preference within ethnic groups is closely linked with skin color is often ignored. Recent research, however, has found that among Blacks, lighter skin has major implications in the job selection process—where one is better off if he/she is lighter-skinned. Due to issues of attractiveness and general levels of comfort, individuals tend to feel a lighter-skinned black is more competent or less threatening, respectively. Though many companies are now concentrating efforts on enhancing diversity—with race being one of the primary focuses—one has to wonder if these "advancements" in diversity are resulting in more lighter-skinned Blacks being hired over their equally-qualified darker-skinned counterparts. This research commentary intends to look broadly at the executive boards of corporate America to investigate if this "lopsidedness" is indeed present. It is expected that greater numbers of light-skinned Blacks will be found in these positions, which will support prior research and illustrate the need for greater discussion and future research regarding this very issue.

In: Anatomy and Physiology Research
Editors: N. R. Zirena and X. B. Liu

ISBN: 978-1-62942-595-8

Chapter 50

Color in Weightlessness Conditions: "μgOrienting" Project

Irene Lia Schlacht[1*], Matthias Rötting[2#] and Melchiorre Masali[3†§]
[1] PhD Student, Chair of Human-Machine Systems,
Dept. of Psychology and Ergonomics, Technische Universität Berlin, Germany
[2] Chair of Human-Machine Systems, Dept. of Psychology and Ergonomics,
Technische Universität Berlin, Germany
[3] Cattedra di Antropologia, Dipartimento di Biologia animale e dell´uomo,
Università di Torino, Italy

Research Summary

In outer space habitats, where the weightlessness and isolation deeply influence human life, color perception, processing and reaction to color are subjects for analysis in Human Factors investigation. The "μgOrienting" project aims to improve the life quality in outer space by research on colors and other visual stimuli.

Picture 1. Color Design research in International Space Station (ISS), Schlacht 2007.

* isc@mms-tu-Berlin.de www.mms.tu-berlin.de or Irene.Schlacht@gmail.com www.extreme-design.eu) I. L. Schlacht, TU-Berlin, IPA Fach MMS, Sekr. FR 2-7/1, Franklinstr.28-29, D-10587 Berlin
# mro@mms-tu-Berlin.de
† Melchiorre.Masali@gmail.com.

In: Anatomy and Physiology Research
Editors: N. R. Zirena and X. B. Liu
ISBN: 978-1-62942-595-8

Chapter 51

WIUD Experiment: Colors and Visual Stimuli for Outer Space Habitability

Irene Lia Schlacht,[1*] Matthias Rötting[1#] and Melchiorre Masali[2†]
[1]Chair of Human-Machine Systems, Dept. of Psychology and Ergonomics, Technische Universität Berlin, Germany
[2]Cattedra di Antropologia, Dipartimento di Biologia animale e dell´uomo, Università di Torino, Italy

Research Summary

In microgravity under weightlessness conditions, where 'Up' and 'Down' have no meaning, orientation is of primary importance. Instinctual reactions to color and symbols are investigated in the WIUD experiment to help implement Up and Down orientation in Outer Space Habitats.

Image 1. Expedition 9, Astronaut Mike Fincke juggles oranges in zero gravity. (c) NASA.

* isc@mms-tu-Berlin.de www.mms.tu-berlin.de or Irene.Schlacht@gmail.com; www.extreme-design.eu; I. L. Schlacht, TU-Berlin, IPA Fach MMS, Sekr. FR 2-7/1, Franklinstr.28-29, D-10587 Berlin.
# mro@mms-tu-Berlin.de.
† Melchiorre.Masali@gmail.com.